ANIMAL MUSIC:
SOUND AND SONG IN THE
NATURAL WORLD

BY TOBIAS FISCHER & LARA C CORY
CONTRIBUTIONS BY KATE CARR & SLAVEK KWI
EDITED BY GAËLLE KREENS

First published by Strange Attractor Press 2015
© The Authors 2015
ISBN 978-1-907222-34-4

A CIP catalogue record for this book is available from the British Library.

Strange Attractor Press
BM SAP, London, WC1N 3XX, UK
www.strangeattractor.co.uk

Printed in the UK

Cover & CD image by Travis Bone
Layout and design Emerald Mosley

Printed and bound in Great Britain by
TJ International Ltd, Padstow, Cornwall

All chapters by Tobias Fischer (TF) and Lara C Cory (LC) unless otherwise stated

PRESENT IN THE WORLD

INTRODUCTION: A QUEST FOR WONDER, A QUEST FOR KNOWLEDGE

All across the world, artists and scientists are listening to the sounds and songs of animals. What, exactly, are they looking for?

Well into the 20th century, Costa Rica pursued a deforestation policy that plundered some of its greatest natural resources. Then, in an unexpected turnaround, a full thirty per cent of the country's richest ecosystems were declared national parks and wildlife sanctuaries – more than anywhere else on the planet. After 40 years of careful restoration, it is now a leader in the field of ecologically friendly tourism. The Corcovado forests at the heart of the Osa peninsula in particular are considered to be a national treasure and one of the most impressive areas on the planet. Condensed into a space of just 425 km², the reserve offers dense greens, myriads of fascinating animal species, towering trees, and, according to Costa Rican natural history editor Daniel Janzen, 'the complete tropical insect ecosystem from Mexico to Panama'. Here, alone in 'the most biologically intense place on earth', French field recordist and sound artist Rodolphe Alexis is using his ears to take in the beauty surrounding him, travelling the woods as part of a multichannel recording project about forest ecosystems. Alexis was fascinated by the idea of animal symbolism, of toucans, macaws and tapirs representing 'wildlife' and 'the new world'.

Despite his already high anticipation, the actual sensation of experiencing the astounding richness of this music of the forest far surpassed his wildest dreams. For an entire month, Alexis roamed the park for interesting spots to record. His explorations mostly commenced after dusk, using his feet, boats and cross-country vehicles to move from one place to another. And yet, the richness of the best sound locations is most clearly expressed much later, at dawn. One by one, in waves of gradually intensifying sounds, animals are waking from their slumber, their voices building to a stupefying climax, then slowly subsiding again, as they commence their daily activities. Then at night, the goosebumps return: 'One is struck by the originality and diversity of sounds, the very large number and abundance of sources'. Alexis says, 'Amphibians and insects take the lead. It is overwhelming and confusing.'

Titled *Sempervirent* and originally conceptualised for a surround sound installation, his project is, on paper, an audiovisual one. Still, to someone like Alexis, the phonography of a site is always the most engaging part. Recording, as he puts it, 'becomes an attitude, a poetic relationship with the fleeting sensation of being "present in the world" while creating a distance with reality through the filter of the microphones and headphones' And through this distance, what formerly seemed to be nothing but a chorus of blissful noise is suddenly revealing patterns, a structure, a meaning: 'It seems that animals have an aesthetic conscience, a judgement. They can be sensitive to certain "attractive" aspects of artefacts, and aesthetic criteria guide their choices.'

In the recordings for *Sempervirent,* which was easily one of the most celebrated field recording albums of 2012, Alexis gets as close to his subjects as possible to arrive at answers. The same desire is driving his colleague Jez riley French, as he sets out to capture the sounds of ants munching away at fruit in a small Italian village. Like Alexis, French isn't that interested in approaching animal noises from a scientific angle. Rather, what motivates him is the sound itself, its personal character, its inexplicable nature: 'When I record animals, I'm attracted to the moments spent listening to them in situ rather than in collecting species recordings. The more time I spend recording animals, the more I think about what it is that their sounds create.'

Using strung contact microphones underneath and inside a windfall apricot, he can spend hours immersing himself in a cosmos made up of all but inaudible sounds, of emanations so quiet they would hardly register within the bigger picture. Underneath French's headphones, though, they are anything but insignificant, creating a small but intricate sound world, which he gladly shares with interested passers-by. Of course, there is a difference between these eating noises and the communication signals from the ants. But French doesn't want to get too caught up in technicalities: 'I'm not too interested in knowing the species names of creatures I record. Not because I don't value that particular approach, but it's simply that I'm not a documentary recordist. I guess there's a balance between knowledge of what one is hearing and accepting that what we know is slim at best. Listening has different forms and so concentrating on the language of animals is just one way.'

Both Alexis and French consider themselves sound artists rather than documentarists and their albums tend to be carefully conceptualised, edited and composed. In a way, however, they represent different ends on the scale of arriving at a better understanding of the significance of animal song. From the earliest days of man, the relationship with other species has been as much about survival as it has been about philosophical and social questions: Clearly, there is a communicative aspect in animal expressions, but does that mean they can be considered language? There is a creative aspect to animal song, but can it be considered music? And if we can understand the meaning of these expressions, can we start to truly engage in a two-way dialogue? Answers to these mysteries can be found in treatises ranging from the ancient Greece to the 18th century writings of German philosopher Leibniz, on all continents and in all cultures. After thousands of years of pondering and analysis, a final verdict is still remote. And yet, the quest has possibly become more urgent than ever. By discovering what we share with nature and what sets us apart from it, we are arriving at a better understanding of who we are ourselves – and what we can do to restore that bond forged over millennia of evolution and development.

---

http://costa-rica-guide.com/Natural/Corcovado.html

MUSIC OF THE PLANTS, MUSIC OF THE ANIMALS

*Shamanism offers fascinating possibilities of seeing – and hearing –
the world like animals.*

Sound is power for the shamans of the Amazon basin. The magical melodies, called icaros, taught to these healers by the plants and animals of the jungle bring with them not just the power to heal, but the power to kill, to protect, to find love, get rich and even to uncover the infidelities of a wandering spouse or partner.

A visitor to the shamans of the mestizo peoples of the upper Amazon may come seeking to uncover something as prosaic as why his or her business failed, but they are equally likely to be seeking a remedy for a physical illness or an emotional cure for a malaise of the soul.

Shamanism in the Amazon basin is not a long faded ancient practice, but a contemporary reality in indigenous and mestizo communities of the region. Although no doubt a varied and dynamic tradition, shamanism in its many forms presents a vivid example of a way of life structured by a belief in the power of sound and song. It is the magical melodies, the icaros of the plants and animals, which heal the sick, protect the vulnerable, uncover the lost and divulge answers to the many questions brought to the shaman. And it is the shaman's job to learn these songs.

Learning to sing

In his astonishing book *Singing to the Plants* Stephan V Beyer asserts that the primary task of the shaman is to sing. Beyer's book presents an overview of the shamanic traditions of the upper Amazon, but it also fascinatingly documents the author's own attempts to train as a shaman. This training involves entering into a complex apprenticeship with both a shaman teacher (referred to as maestro ayahuasquero) but also the plant spirits themselves. Both of these relationships revolve around the ingestion of one plant in particular, the ayahuasca vine.

Meaning 'vine of the soul' in Quechua, ayahuasca is a psychoactive plant which induces visual and auditory hallucinations, and in many cases violent vomiting. Beyer says that by taking the plant into the body, the shaman is able to visit with the spirit of the vine who will teach him or her the plant's icaro. 'To learn the plants', Beyer writes, 'means to create a relationship with the plant spirits, by taking them into the body, listening to them speak in the language of plants, and receiving their gifts of power and song.'

This, however, is no simple process – plants are fickle and do not readily give up their secrets. To convince them, the shaman must go to extraordinary lengths to show he or she is worthy of possessing the magic melodies. Because plants are seen as extremely jealous, shamans must abstain from sex. The plants also hate strong smells, particularly those associated with sex like semen, but also menstruation.

The shaman must also purify his or her body by following a highly restrictive diet, avoiding any strong tasting foods as well as sugar and salt. Some shamans believe they must avoid sunlight. Others, like the shaman Pablo Amaringo, believe that because the spirits generally dislike the smell of humans it is best for shamans to avoid other people and spend as much time as possible in the jungle ahead of an ayahuasca ceremony.

Swallowing knowledge

Ingestion remains central to shamanic study. After learning the icaro of the ayahuasca vine, shamans ingest other plants in order to study them. This process, Beyer states, is viewed as spending time with the different plants and learning from them. Other substances are also studied in this way. To study steel and learn its icaro the shaman soaks it in water for several days and then drinks the water; gasoline is studied by inhaling it. Cologne is mixed with ayahuasca to access its properties and learn love icaros.

During these sessions with the plants, animal spirits may also appear to the shaman to teach their icaros. Anthropologist and ayahuasca researcher Luis Eduardo Luna says one shaman he spoke to, don Manuel Córdova-Rios, brings animals to his students through imitation, enabling them to 'become one' with the animal, and in time use their icaros to call on the animal spirit for help in healing ceremonies.

This way of learning presents a startling reversal of the Western preoccupation with the importance of observation. For the shaman the body is the site where knowledge is produced, and the way it is transmitted is via sound and melody. 'In fact it seems that one of the central ideas is that certain qualities or properties of plants, animals, minerals, or metals can be incorporated either by the ingestion of some part of this object, or by other means unknown,' Luna writes.

Mastering the icaro of the plant or substance in question is the ultimate aim of this process of ingestion. According to Luna, for many shamans these songs are seen to contain the very essence of the plant or animal, and can even be a substitute for the physical substance in the case of medicinal plants.

Songs of power

Discussing one of the ayahuasca ceremonies held by his teacher don Roberto Acho Jurama, Beyer describes don Roberto beginning the session by singing a simple song used to protect those in attendance. He then sings the icaro de ayahuasca in a whisper over the ayahuasca liquid and distributes it to those attending, before singing again and drinking it himself. As the room fills with the sounds of retching and vomiting, Beyer says don Roberto calls the plant spirits, again by singing, reciting the icaros of a vast number of spirits to the healing room so they are ready at hand for the ceremonies to come.

Despite having consumed ayahuasca in the past, Beyer writes of his apprehension at swallowing the 'vile liquid'. 'It is one of the worst things I have ever tasted,' Beyer writes, 'it coats my teeth and tongue.' Before the ceremony, Beyer says he wonders if the spirits will appear to him tonight.

In a book chapter devoted to the icaros of the Peruvian Amazon, Luna details some of the examples he encountered in the Amazonian provinces of Peru.

Despite the fact that it is plants which are seen as the original teachers for the shaman, there are icaros of both plants and animals. 'A man is like a tree. Under the appropriate conditions he grows branches. These branches are the icaros,' don Alejandro tells Luna. An icaro of the sloth, an animal regarded as clean, strong and very picky about what it eats, is used to cure digestive problems in children. The flamboyant toucan, with its sad and beautiful song, has an icaro which is used to help win love. Icaros of slippery animals are used to aid in child birth. Other animals like the eagle, the condor, the boa, the eel and the jaguar are all called on by shamans for healing.

As Luna writes, 'Through the icaros the shaman is able to "become one" with the animal and see the world accordingly.'

At times an animal, rather than a plant, may even become the primary teacher of the shaman. Luna cites one young shaman who said his teacher was a hawk. Another shaman said during an intense period of ayahuasca ingestion and learning he had felt himself becoming very small, and subsequently a large ant had appeared to him and communicated in a 'tridimensional' language.

One example of an icaro of protection cited by Luna was used by the former shaman don Pablo Amaringo, which invokes the spirit of the jaguar.

Part of the icaro goes: 'Where are you coming from / the offspring of the black jaguar / You nourish the earth with the milk of your breasts / In this way come forth. Behind it comes / the otorongo [jaguar] is calling him / in the midst of the great forest / it comes screaming.' But icaros don't just invoke the characteristics of the animal or plant in question, they also reflect the realities of contemporary life. One is used to invoke an insect which has eyes similar to the headlights of a car. This icaro, Luna says, is used to help the shaman look for a person who has been stolen. Some icaros are used simply to farewell a good person, or attract a particular fish prized for its taste.

Each shaman has a 'main icaro' which represents the essence of his or her power. If a shaman learns the main icaro of a different healer he or she will inherit the knowledge of that healer on his or her death.

Icaros can be whispered, whistled or sung in full voice. Some have clearly discernible words, others are in a mix of languages, some are in the languages of the spirits and some are just a melody.

Links to the past

The importance of sound to shamanism may find its roots in some of the traditional beliefs of indigenous communities in the Amazon. One striking example is found in the myths of the Wakuénai, described by anthropologist Jonathan D Hill in an article documenting the practices of the indigenous curers of the upper Rio Negro basin of Venezuela. According to Hill, in the mythology and cosmology of the Wakuénai, whose ancestral lands are found around the Isana and Guainia rivers in Venezuela, the world as we know it was literally sung and hummed into existence.

For the Wakuénai, the figure of Kuwái, the son of the world's first man (Iñápirríkuli) and woman (Amáru), produces a 'powerful sound' which 'opens up' the world, transforming it from its past existence as the miniature world of his parents, into the life-sized world inhabited by the humans, animals and plants of today. Kuwái is subsequently killed by his father and the world shrinks back into its miniature form, emphasising the integral link between sound and human existence.

Again it is through music that the world is reconstituted. From Kuwái's ashes grow plants which are used to make the sacred flutes and trumpets played in initiation rituals and sacred ceremonies. The instruments are stolen from Iñápirríkuli by Amáru who, along with other women, play them as they lead him on a long chase across the world, reopening it for a second and final time in its current form. Hill traces the origins of shamanic practice in this region to these beliefs of the Wakuénai, asserting that the most striking feature of the healing rituals in the region is the degree to which the behaviour of the ritual specialists centres on musical performance.

'The rituals are essentially musical events around which a variety of other less important activities revolve,' he writes.

In re-enacting Kuwái's musical creation of the world, Hill states the chant owners tap into the transcendent power of sound and song.

A power beyond words

Words may be crucial to the recitation of many icaros, but their intelligibility is not, and in many cases meaning is actively obscured. For Beyer's teacher don Roberto, the more abstract, less conceptual, less overtly intelligible the icaro, the more powerful it is. He states that many of the most powerful icaros feature a mix of languages, some unknown, some indigenous, other gleaned from the plant teachers, as well as non-verbal sounds.

When reciting icaros, shamans add to the indiscernibility of the songs by dissolving many of the words into whispers, and non-verbal sounds. Other icaros have no words at all, consisting solely of a melody.

Ayahuasca seems to powerfully affect the conception those who ingest it have of their ability to understand the meaning of pure sound, or non-verbal sounds. In *Singing to the Plants* Beyer cites the experience of anthropologist Janet Siskind who stated that after ingesting ayahuasca she believed she was capable of understanding every song she heard.

Doña María Luisa Tuesta Flores, another of Beyer's teachers and one of the few women shaman referred to in the pertinent articles, said she spoke to the spirits in Quechua, despite not being able to speak the language when not under the influence of ayahausca. And both don Roberto and doña Maria said spirits from out-of-space spoke to them in a 'computer language' of beeps, while the languages of the plants themselves were described by the pair as consisting of somewhat similar high pitched sounds. According to Beyer, both shamans report no difficulty in understanding what is being communicated to them in these languages.

Given the shamans tradition's emphasis on sound, the power many of these healers' believe exists in its absence – silence – is surprising.

According to Beyer, the most powerful of all sounds within shamanism are those which hang between song and silence. And the most potent icaros are often refined into breathy whistles and inaudible whispers.

In these instances sound itself appears as a readily discernible language, its power being bolstered by its faltering existence at the edges of perception.

This is perhaps most profoundly summed up in Beyer's claim that for the shaman, the most powerful sound of all, one which can both kill and cure, is the one which most nears silence: The sound of blowing, a soft, voiceless 'pshoo' ...

Stephan V Beyer, Singing to the Plants - A Guide to Mestizo Shamanism in the Upper Amazon, University of New Mexico Press: Albuquerque, 2009

Jonathan D Hill, 'A musical aesthetic of ritual curing in the northwest Amazon' in Portals of Power, University of New Mexico Press: Albuquerque, 1992

Luis Eduardo Luna, 'Icaros, magic melodies among the Mestizo Shamans of the Peruvian Amazon' in Portals of Power, University of New Mexico Press: Albuquerque, 1992

IMITATION AND INSPIRATION: OLIVIER MESSIAEN

*Animals have inspired composers for centuries. In the work of
Olivier Messiaen, however, birds are more than a mere influence –
they're co-creators of the music.*

A maximalist in every meaning of the word, yet seemingly disinterested in the modernist race for progress, Olivier Messiaen was one of the 20th century's most singular artistic characters. Unlike the majority of composers of his time, he neither tried to anticipate the future in his pieces (as Igor Stravinsky had done in his youth), nor did he seek to revitalise the present by means of harking back to the past (as Stravinsky would later do in his controversial neo-classical phase). Instead, he wrote for the present and according to his own, personal convictions and his faith alone. Not only did this allow him to avoid getting caught up in the great ideological battles raging around him which brought as many inspiring ideas as created victims. It also meant he didn't have to turn his back on his audience or defend his work through intellectual, yet ultimately artificial justifications. As a remarkable exception among modern composers, Messiaen gained not just notoriety and professional recognition during his lifetime, but true admiration and what one might today call 'a solid fan base'. This, in turn, would lead to a string of high-profile commissions which, in combination with his job as a professor at the Paris Conservatoire, provided him with the financial independence required for avoiding compromise. And as the eccentric, humorous and yet deeply serious personality that he was, Messiaen made full use of this freedom to create the arguably most idiosyncratic and identifiable body of work produced by any composer in recent times.

Over the years, Messiaen has become known even among lay audiences for a variety of particularities of his compositions and approach towards writing – the latter meticulously laid out in his brief, but extremely dense 1944 *Technique de mon langage musical* (Technique of my musical language). Long before the American minimalists, he would seek inspiration in the rhythms of Africa and India. He would inspire the entire serialist movement with a single, short piano piece titled 'Mode de valeurs et d'intensités'. And his continuing insistence on the use of 'musical colours' would provide a creative stimulus to a new generation of artists, directing their attention more towards timbre than form. And yet, all of these seemingly disparate ideas are conclusively held together by his interest in bird song, to a degree that led French documentary director Olivier Mille to claim that 'within his oeuvre, the bird takes centre stage'. This statement is definitely up for debate. What can be said with absolute certainty, however, is that this deep and life-long fascination was never a novelty. Rather, it seemed to allow the composer to arrive at a more profound understanding of his faith, his art, the world around him and his own purpose in it. As such, ornithology was part of a much larger philosophy of life for Messiaen and permeated every aspect of his life.

Without precedence

It can safely be asserted, too, that Messiaen's approach to working with bird song as musical material knew no precedence in Western art music. In her astutely argued and comprehensive essay 'Crickets in the Concert Hall: A History of Animals in Western Music', Canadian composer Emily Doolittle researches the question of how animal song has been used in compositions from the earliest times of written notes to the present time. What is remarkable about her study is how long it has taken – roughly 500 years to be precise – for these inclusions to grow from being a mere quip or musical joke, as in Mozart's KV 522, purportedly written as an homage to his deceased starling, or a trivial rhythmical motive, as with Haydn's 'The Quail', to a true symbiosis. As Doolittle asserts, until the romantic age, neither did the musical aspect of animal song penetrate beyond the surface of the music, nor did the composers show any outward interest in tapping into the rich folk tradition, which had gradually perfected imitative techniques to create a dialogue with animals. This is not to say that there have not been particularly striking uses over the centuries. Acclaimed pianist Matthew Schellhorn has even built an entire, feature-length programme around the theme of bird song, uncovering a wealth of material from different ages: 'It is amazing to think that birdsong has been a constant inspiration to composers for hundreds of years! What is remarkable is the variety of ways in which birdsong can be woven through the music. So, we have in Louis-Claude Daquin's piece "Le coucou" an almost ceaseless thread of the cuckoo's song, forming a backbone to the tonal and harmonic content; in Rameau and Ravel we have what might be described as "generic" song infused into a complex musical creation that explores natural fluctuations of mood. Birdsong, and the subject of the bird, is often a "way in", therefore, to a deeper understanding of reality.'

After centuries of relatively little interest in animal song, it became all the rage during the romantic age. Examples of pieces either featuring animal imitation, or at least allusions to them, quickly multiplied, to an extent that Gustav Mahler developed an outright aversion to it and, as indicated by Paul Schiavo, came to prefer deeper and more subtle ways of creatively including nature in his work: 'Mahler's music avoids (...) tone painting in favor of a more encompassing view of its subject. Instead of depicting nature through aural mimicry of wind and water and animals, Mahler conveys what might best be described as the soul of nature – or, perhaps more accurately, the response of his own soul to nature.' It was a response shaped by pieces like Beethoven's 'Pastoral' Symphony, with its second movement, titled 'By the Brook', whose conclusion contained references to nightingale, quail and cuckoo, as well as Camille Saint-Saëns' 'Carnival of the Animals', a suite originally written for private performances, which rapidly gained fame after his death and would turn both into a concert staple and a popular educational introduction into the world of classical music in the 20th century. Although these works have lost nothing of their compositional power, the link to the world of animals is a playful one at best. Animal calls are abstracted and subsumed into a human art cosmos, in which they serve purely functional purposes and are subsumed to the thematic development taking place around them.

Fundamental changes

It would take until the mid-20th century before not just the compositional approach to animal song, but the general perspective on it, would undergo fundamental changes. Messiaen may have been a pioneer and a harbinger of change, but his ideas were certainly embedded into a far more wide-reaching transformation process. This gave his compositions an immediate social relevance. As Doolittle writes: 'In the past half-century or so (...) a variety of scientific, social, and philosophical trends have combined to change the way animals are regarded in the West, which in turn has altered the way animal songs are treated in Western music. Among these changes would be the lessening influence of the Judeo-Christian world-view, as secular public institutions replace religious institutions; increased recognition of animal rights, brought to public attention by such writers as philosopher Peter Singer, animal-scientist Temple Grandin, and primatologist Jane Goodall; recent discoveries of the extent to which animals may use tools and language; recognition of self-awareness, cognition, emotion, and personalities in some species; and increased awareness of our interconnection with the natural world, in face of an impending ecological crisis. Whereas it was previously possible for Westerners to regard animals as completely "other" than humans, we are now increasingly inclined to emphasize similarities between animals and ourselves.'

And yet, there is something else about Messiaen's ornithological explorations that makes them stand out in the by-now voluminous repertoire of bird-related music: There is neither an attempt at expertly crafted imitation, as had been the case with some of the composers mentioned in the previous paragraphs. Nor was he part of an effort of faithfully mapping the acoustic world around us with new, portable recording devices. Instead, Messiaen's bird pieces exist in a space between the purely musical and the factual, working both with imitative and creative devices to delineate a unique and surpassingly imaginative music – or, as he once put it himself: 'I'm writing something that isn't real, but plausible'. Messiaen's biographer Peter Hill, a renowned pianist in his own right, has compared this approach to that of a painter going out into the open with his easel and using the sounds of nature as his palette. What it also means, of course, is that, in his oeuvre, bird song and human composition techniques are merged to a degree that reveals fundamental insights on what, precisely, is unmistakably and originally human about music, and what may have been appropriated later.

Development as a composer

Although Messiaen began showing an interest in birds as a boy and wrote a short bird-inspired piece, 'La colombe' (The Dove), as early as 1929, it would take until the mid- to late 1950s, when the composer was already approaching his 50th birthday, before he began explicitly including them in his publications. The one notable exception to that statement is his 'Quartet for the end of time', which he wrote between 1940 and 1941 as a prisoner of war and which contained parts 'marked "comme un oiseau" (like a bird) (...) according to the preface to the score, the clarinet is a blackbird, the violin a nightingale, two birds that sing at dawn, thus adding another level of metaphor to the musical message.' The 'Quator', despite (or precisely because of) the sad facts surrounding its genesis, remains a firm concert staple and would, at the time, lay the foundations for his academic career and renown as an artist. And yet, it was only a decade later and probably thanks to his studies with ornithologist Jacques Delamain, that he began turning towards bird song more systematically, eventually developing a curious blend of scientific ardour and child-like enthusiasm.

Messiaen would visit the countryside for entire days, leaving the house in the early morning to capture the rising sun and coming home only after it had gone down again, scribbling everything that happened in between into myriads of *cahiers* (notebooks), which would eventually add up to thousands of pages of observations jotted down in all but illegible sentences of haiku-like brevity. There are detailed descriptions of how he would hurry through the field, occasionally taking his pupil Yvonne Loriod with him, only to spend hours at the same place to fully take in its acoustic potential. In Mill's biographical documentary *La liturgie de cristal* the composer can be seen standing in the middle of the woods with a pen and a piece of paper, listening attentively and describing his perceptions to the camera. His voice is firm and authoritative as always, but his eyes light up like those of a boy speaking about his favourite cartoon series. And to this day, Peter Hill remembers the day that he handed the composer a book about the birds of England as a present: For ten minutes, Messiaen just sat in silence, flicking through the pages and murmuring quietly to himself whenever he recognised something.

This obsession with birds has presented generations of musicologists with a riddle: What, precisely, was he looking for? Messiaen himself loved evading the question, comparing his interest in ornithology to his vocation for music in general or to love – to aspects of life, therefore, for which there could and need not be any rational explanation. Only occasionally, unlikely perhaps for an otherwise highly emotional and impulsive character, he expressed his admiration for bird song as being related to the birds' unbridled creativity, to them inventing anything from the chromatic and diatonic scale to quarter tones, sixth tones and even group improvisation. To Hill, on the other hand, it seems clear that to the faithful Catholic Messiaen, birds were always religious symbols as well – one need only look at the importance of the dove as a representation of the Holy Trinity – and that the natural space they inhabited was a manifestation of God's glory on earth. It is a position shared by German composer Theo Hirsbrunner, who visited his colleague several times in his Paris apartment and, after many private conversations, arrived at the conclusion that, for Messiaen, the goal of using birds as a departure point for music, was to 'present their song and the sounds of nature in a hallucinatory light'.

Emotional turning point

In a way, Messiaen's music marks an important turning point for the arts in that depicting the beauty, wonder and greater truth concealed in and expressed by nature suddenly became possible again. Japanese pianist Momo Kodama, herself one of the leading Messiaen-interpreters of current times, describes it thus: 'After the middle of the 20th century, composers had the tendency to move away from music inspired by and describing nature, character or sentiments and towards a more abstract expression. Messiaen came back to nature. Birds play the main role, but also the surrounding nature such as the sea, the trees, the frogs, the pond and more are described through complex and virtuosic writing.'

It can hardly come as a surprise, then, that his bird-period falls precisely into an extremely intense phase of his career. On the one hand, he experienced the biggest triumphs of his life, with the 'Turangalîla-Symphonie,' a grandiose orchestral work, receiving rapturous acclaim and cementing his fame in the USA. On the other, his private life suffered from his first wife's mental illness, his growing feelings towards his pupil and congenial interpreter Loriod, as well as the immense stress of dealing with a surplus of responsibilities – ranging from his commissions, housework and raising his son to his duties as an organist and, of course, constant studies in the field. His diary from the time is filled with feelings of depression, fear and doubt as well as an increasing sense of no longer being able to make his own decisions. It is perhaps precisely for this reason that he began to appreciate birds for their absolute clarity and a sovereignty he couldn't arrive at in his own life: 'In melancholic moments, when my own uselessness is becoming painfully clear to me, when every musical language – be it classical, exotic, old, modern or ultra-modern – seems to me no more than the praiseworthy result of patient research, without any kind of meaning behind the notes which might justify the amount of work invested into it, what else can one do but search for the true countenance of nature, which is resting forgotten somewhere in the forest, in the field, in the mountains, at the beach, among the birds? (...) Rilke once wrote the magical words: "Music: Breath of statues, maybe: Silence of paintings. You language, where all language ends ..." Birdsong exists on a level beyond the dream of this poet. It certainly exists beyond the composer trying to write it down!'

Trying to write it down would nonetheless turn into a personal quest. For the Donaueschinger Musiktage in 1953, Messiaen scored a piece for piano and woodwinds ('Réveil des oiseaux') aiming to recreate a dense panorama of bird song, following it up with 'Oiseaux exotiques' for piano and orchestra in 1955-56. The pieces met with mixed reactions, with some praising their innovative character and others failing to see the point. Eventually, they even led to a rift with his publisher. But it was only in 'Catalogue d'oiseaux', an epic, three-hour long cycle, that his efforts truly came to fruition, a work which Kodama describes as 'the culmination of Messiaen's research as an ornithologist and the development of his harmonic and rhythmic processes.' What made it stand out was not just the increasing ease with which Messiaen transcribed complex bird sounds to the staves and the growing confidence with which he integrated them into his own creative cosmos. They are also testimony to his determination and will to break down the borders between the world of art and real life. For 'Oiseaux exotiques', he had still worked mainly with bird recordings from the box set *'American Bird Songs'* by Comstock Publishing as a basis for his transcriptions. For the 'Catalogue d'oiseaux', however, he exclusively made use of notes taken on his trips into the French countryside. Originally conceptualised as a modular work including all major birds in France, it was also easily the most scientific project he'd ever taken on, based on his own research and a wealth of literature. And although he would never come to see his initial plans through in full, the result would nonetheless constitute a work of intimidating proportions and impressively ambitious scope.

Ambitious and demanding

To understand just how ambitious and demanding the 'Catalogue' really is, you only need to speak to the select group of performers who have either performed or recorded the composition in its entirety. To this day, there are only a handful of interpretations available, from Yvonne Loriod's rendition in close temporal proximity to the piece's debut – published on Erato Records and long out of print – through Anatol Ugorski's recording for Deutsche Grammophon and Martin Zehn's box set for Arte Nova Classics. Even Pierre-Laurent Aimard, considered by some as the ideal Messiaen performer, has only published extracts of the catalogue. With this scarcity of available interpretations in mind, each new release is sure to attract plenty of attention – which only makes the task all the more intimidating. And yet, a new generation of musicians has taken on the challenge and made the piece their own. Kodama and Schellhorn are among them. Schellhorn has studied with Peter Hill, received the personal blessings of Yvonne Loriod shortly before her death in 2010, and has received great acclaim for a disc of Messiaen's chamber music.

Kodama seemingly first spent almost two decades living and working with the music before presenting her own take on it. The depth of her insights into the piece has only served to increase her reverence: 'The big challenge and difficulty was to try and understand the pulse of the piece – not a pulse in the sense of a heartbeat or the human voice, to which we are used in classical music. But a pulse of nature which is much more free, but also has a special and unique logic. Unlike a human being, a bird does not take a big breath before starting to sing and can stop singing suddenly and abruptly – the rhythm of the "conversation" between birds is very different according to the different species – and the most difficult technically to play is when one "chattering" bird is played by the right hand, and simultaneously on the left hand another bird is singing as fast but with a completely different rhythm. This is one of the hardest things to play in the whole piano repertoire, but something which has a great effect.'

Precisely because it is so difficult, some of the catalogue's most ardent admirers have gone to great lengths to get it right. Even Yvonne Loriod initially met with her future husband's disapproval, who considered that her take on the 'Réveil d'oiseaux' was too mechanical and did not create the right timbral colours. It was only after she took a trip into the woods at Orgeval with her mother to witness the dawn chorus that her interpretation gained the necessary hands-on quality. Schellhorn, too, is convinced of the necessity of learning as much about bird song as possible, both theoretically and practically, to arrive at an interpretation which bridges 'the divide between the natural and the unnatural'. To accomplish this daring tightrope walk, he has studied books and CDs on the subject, experimented with mimicking birdsong on the piano and even literally followed in the composer's footsteps: 'I have visited the areas of France that inspired the 'Catalogue d'oiseaux' and also 'La Fauvette' – the Meije Glacier and the Dauphiné (Isère), for instance. Here, I have got a feel for the landscape; I have seen – and heard! – the environment that has inspired this music. Perhaps I have heard some descendants of the birds Messiaen himself heard.' It is an approach which Messiaen, who once visited the lands between Massada and the Dead Sea to hear the birds Jesus might have heard for one of his pieces, would have appreciated.

Schellhorn's programme around the theme of bird song, which encompasses some of the earliest known examples, such as Louis-Claude Daquin's 'Le coucou' from 1735, as well as a recent work by Schellhorn's friend and composer David Bruce, which 'explores a poem by Wallace Stevens where a bird inhabits our world not only on the physical level but also the psychological', has provided him with some clues as to why the connection with birds may have been so inspiring both for human kind in general and Messiaen in particular: 'The bird is an ever-present part of our human experience. It is also universally identifiable with freedom, with simplicity, with beauty. Birds have fascinated man from prehistory and they have developed symbolic and religious significance through the ages, from the Roman eagle of the Caesars to the dove of the Bible, from the fantastic phoenix to the griffin and the god-peacock of India. In its interaction with man, it is found in every cultural area, in music of course, in visual art, in poetry, even in heraldry. As to the music of birdsong and why we love it, I should say that beautiful singing is enjoyed by humans and birdsong is understood as the most natural manifestation of song. We are interested by its familiarity and its unpredictability, its contrasts and its consistency.' While Messiaen found birds to be 'sovereign' in their creative capacity, he also said they are 'the closest to us, and the easiest to reproduce'. 'I should assert that the only man-made music ever, perhaps, to come close to birdsong is Gregorian chant. This music, the music proper to the liturgy of the Roman Catholic Church, manifests the flexibility of both melody and rhythm. There is even evidence to suggest that the melodies we have written down were the basis, in fact, of improvisation – which of course reminds us of the sounds of the natural world.'

Practical reasons

Simply put, practical reasons may have governed our love of bird song and Messiaen's choice of birds as the subjects of his compositions in the 1950s and beyond. It is for the same practical reasons, next to his obvious desire to work with Yvonne Loriod as much as possible, that he chose the piano as his main instrument with which to translate bird song into the domain of human hearing and the realms of human art: The piano was the only instrument with the required gamut and immediacy of attack to allow for the kind of rapid, high-frequency signals produced by birds. It is also the only instrument capable of imitating, partly by means of unusual tonal combinations, the unique, recognisable timbres of birds like crows or sea gulls. This encourages the interpretation that the influence of animal song on our music is one shaped foremost by familiarity, empathy and, ultimately, constant transformation. In composing the 'Catalogue d'oiseaux', Messiaen was, as it were, placing this process of inspiration in the limelight, open for everyone to see. Rather than imitating birds perfectly, he worked with them and rather than hiding their influence, he fully admitted to it. The result speaks of a relationship with nature, which considers both our individual place in it as well as our part of the whole and which still feels natural. Which is probably why this music has remained so utterly contemporary for more than half a century.

It is quite revealing that the first volume of the 'Catalogue' would never be followed up by further episodes, although the composer claimed to have amassed more than enough material. There are plenty of reasons which could account for this, including a severe illness. But perhaps, subconsciously, Messiaen knew that by leaving his work 'unfinished' and 'incomplete', he was adding to its artistic impact. Rather than a brilliantly realised yet ultimately dry scientific study, it would forever remain the fantastical product of his own imagination, which in turn was the product of millions of years of evolution and cohabitation with the animals he loved the most.

Emily Doolittle, 'Crickets in the Concert Hall: A History of animals in Western music' in TRANS / Transcultural Music Review 12, 2008

Richard Schiavo, Mahler's Third Symphony Program Notes for the Seattle Symphony

Richard Taruskin, Music in the Early Twentieth Century, Oxford University Press, 2010

Olivier Mille, La liturgie de cristal, Juxtapositions, 2007

Theo Hirsbrunner, Olivier Messiaen / Leben und Werk, Laaber-Verlag, Laaber, 1999

Peter Hill & Nigel Simeone, Messiaen, Yale University Press, 2005

Interview with Matthew Schellhorn http://15questions.net/interview/interview-matthew-schellhorn

ZOOMUSICOLOGY: WHEN SOUND BECOMES SONG

Most still consider spirituality and music the privilege of humans.
They may well be wrong.

Before humans learned to stretch animal gut over a frame and before they realised the resonant properties of a tightly pulled string, they made music with their bodies; whistling, clapping, singing and clicking. Most likely, early humans mimicked the sounds they heard in the world, from the rhythmic chirp of crickets to the lyrical song of the birds. Studies into the origins of human music suggest many reasons for its development, including protection, spirituality, seduction and important emotional forms of communication. However, it is not until quite recently that we have turned our attention to exploring the reasons and meaning of musicality in animals.

Since our earliest history we've marvelled at the beauty of nature, 'Imitating with the mouth the fluid voices of birds came long before men were able to harmonize light melodies and please the ear.' (Lucretius (94-55BC) *De Rerum Natura*). Composers throughout history, from Handel and Mozart to Delius and Messiaen, have written music inspired by bird song. It wasn't until 1983 that French composer François-Bernard Mâche began to explore the idea that animals might also make music for reasons more than practical communication. In his book *Music, Myth and Nature* he formalised the concept by coining the word 'zoomusicology'. Since then, terms including cognitive musicology, biomusicology, ecomusicology, evolutionary musicology and ornitho-musicology have come into being and we are now, more than ever, looking for universality in music. We are extending our studies beyond the anthropological level and into the animal world where cognitive understanding comes to the fore and mere biological necessity is transcended. As musicologist and semiotician Professor. Dr. Dario Martinelli explains, 'Zoomusicology suggests that music is a zoological phenomenon, rather than a simply anthropological one' and goes even further, to suggest that the two are not in opposition but are in fact inclusive, because humans are, after all, animals too. This area of study is boundless and has serious implications for how we perceive ourselves, and our place in the world.

Animists have long held the belief that not only animals but also the living and non-living world around us share the privilege of a spiritual existence. These beliefs occur in cultures around the world and appear in Shinto, Serer, Hinduism, Buddhism, Jainism, Pantheism, Paganism, and Neopaganism. However, before we can begin to explore this topic, the question of what defines 'music' becomes important. Professor. Martinelli describes his life's work studying the musicality of animals as 'the aesthetic use of sound communication among animals', suggesting that the word 'music' is dangerous whether discussing it in either human or non-human terms. But if we remember that instead of 'searching for the existence of the signifier, i.e., the word music' we are actually 'searching for the signified, the aesthetic use of sounds, which by convention we call "music"', we can begin to think about the concept of animal music.

How do we define animal music? Can we define animal music in the same terms as human music? What might the animal aesthetic be and can we ever hope to understand it? How does it differ between species? Does it have meaning?

The best way to continue on this un-travelled path is to assume that animal music describes the sounds that animals make that go beyond basic communication. Animals communicate with each other using sound and it is a well-documented and researched area of study. Pops, clicks, coos, screams, rumbles and grunts, animals use a wide range of sounds for purposes such as warning calls, mating songs and predation. Zoomusicology on the other hand, is the study of animal song as an expression that goes deeper than utility and in most cases goes beyond human appraisal. In fact, we have to ignore our anthropocentric tendencies and try to remove our subjective notions, ethnological and cultural perspectives when we look at the phenomena of animal music.

We can start by using what we know about human music to help give some value and definition to animal music. Features like repetition of melody and rhythm seem like a good place to start, but such aesthetic signifiers can change dramatically, even among human cultures, which makes it even more unlikely that we can hope to understand the animal aesthetic. And it is essentially the notion of aesthetics that really begins to define music as something apart from sound. So to broaden the scope, aesthetics here will mean phenomena and perception that is not directly utilitarian or serving an obvious practical purpose. Human music may have originated as purely functional, but ask yourself, why do you like music? Why does it stir our emotions? Why do we sing or dance along to it? Why do musicians play it and composers create it? It serves no function for our survival, biologically speaking. Socially of course it serves many purposes but when it comes down to it, the most important and yet unimportant aspect of music is that it gives us an outlet to emotionally express ourselves. Why does the female gorilla sing softly to herself while she grooms her friend? Why do the gibbons call out in musical response to the rising of the sun?

'In a number of cases among song birds, particularly those in which songs of unusual richness and variety are known, we frequently encounter what appears to be musical "invention." This includes re-arrangement of phrases, both innate and learned, and the invention of really "new" material' (Thorpe[1]).

Kairi Kosk from www.zoosemiotics.helsinki.fi/zm has a broad approach to the concept of music, both human and animal. She suggests that aesthetic and practical values don't always have to be mutually exclusive. We seem to have developed a hierarchy where purely functional expression sits at the bottom of the scale leading to the highest point, something we call 'high culture', where the value of the music, art or form of expression exists only in itself. Kosk insists that this hierarchy is not fixed and that music can be a combination of both the aesthetic and practical. Semiotician Thomas Sebeok agreed that there is a 'possibility that animals perform some of the behaviour patterns we observe because they enjoy the resulting experience, regardless of whether such patterns are adaptive or virtually so, but result in a pleasantly satisfying feeling

---

[1] William Homan Thorpe a British zoologist, ethologist and ornithologist. Thorpe contributed to the growth and acceptance of behavioural biology in Great Britain and pioneered the use of sound spectrography for the detailed analysis of bird song.

on the animals' part.' If humans make music for practical and emotional reasons, then is it such a stretch to think that animals might too?

Orlane Ruiz Alonso, also from www.zoosemiotics.helsinki.fi/zm, prefers the divergent theory that music-making stems from the functional behaviour of the 'loud call'. Alonso examines the long and complex duets performed by monogamous gibbon pairs: 'Duets are mostly sung by mated pairs. Typically, mates combine their repertoire in relatively rigid, more or less precisely timed vocal interactions to produce well-patterned duets. Males of many gibbon species produce one or several distinct types of short phrases that often become gradually more complex as the song bout proceeds. The gibbon's song is very loud and complex with pure, melodious and sonorous tones.' The exact reasons for this unusually intricate and melodic expression are unknown. There is no set paradigm to analyse this sort of behaviour, but its exclusivity to couples suggests its function is a form of pair-bonding. Alonso suggests that for both non-humans and humans, the concept of musicality originates from the functional need to communicate and that it will always be so: 'Loud calls in modern apes and music in modern humans are derived from a common ancestral form of loud call. Loud calls are believed to serve a variety of functions, including territorial advertisement; intergroup intimidation and spacing; announcing the precise locality of specific individuals, food sources, or danger; and strengthening intragroup cohesion.' We use music to define ourselves, Alonso believes. We use it to display and reinforce the unity of a social group in instances as varied as national anthems and military marching to religious hymns and youth genre preferences. With consideration to these two opposing ideas about the value of music, François-Bernard Mâche said simply that animals like to play with sounds, and that perhaps we don't need more of an explanation. Professor Martinelli summarises that 'playing is a justification in itself and music is a game in several respects. Playing and having fun are biological functions.'

There is a shared debate between humans and non-human animals, with some people preferring to believe that music is purely functional, and others certain that it is solely for emotional and spiritual expression. We can be certain that animals, like humans, use music to communicate and that it has a biological function in their lives. But can we ever be certain that they, like us, need music for emotional and creative expression? Humans have evolved to exist beyond function alone; is it so foolish to suggest that animals have as well?

A Whale of a Sonata – Zoomusicology and the Question of Musical Structures, http://see.library.utoronto.ca/SEED/Vol5-1/Martinelli.htm

Of Birds, Whales and other Musicians; An Introduction to Zoomusicology, Dario Martinelli, University of Chicago Press, 2009

'Music, Mind and Evolution', Ian Cross, *Psychology of Music*, 2001, Vol. 29, No. 1, pp 95-102

'Ritualization in Ontogeny', W H Thorpe, *Philosophical Transactions of the Royal Society of London. Series B, Biological Sciences*, Vol. 251, No. 772, 'A Discussion on Ritualization of Behaviour in Animals and Man' (Dec. 29, 1966), pp. 351-358

THE SECRETS OF THE DEEP

*There are serious indications that animals may not just be singing
for evolutionary reasons – but because they enjoy it.*

David Rothenberg is excited. The inter-species musician and composer
has been thinking about the relationship between nature and humanity
for a long time, sometimes receiving contempt and mockery for his
ideas. Now, for the first time, he is being taken more seriously by the
scientific community.

His book, *Survival of the Beautiful*, is the culmination of many years
spent pondering the role of aesthetics in nature and evolution. His other
books have examined this notion specifically by species, for example
*Why Birds Sing: A Journey into the Mystery of Birdsong, Thousand Mile
Song: Whale Music in a Sea of Sound* and his latest offering *Bug Music:
How Insects Gave Us Rhythm and Noise*. But Rothenberg is determined
for his readers to see the bigger picture: 'I'm seeing the aesthetics of
whale song as part of a whole larger story of aesthetics in evolution,
which was important for Darwin and is now coming back into biology
through the work of Richard Prum on sexual selection.'

Since the groundbreaking discovery of whale song in 1967 by Roger
Payne and Scott McVay, researchers have compared whale song and its
complexities to human music. Particular elements like its predictable
structure and repetition are the strongest indications that the sounds
are indeed song-like. Other similarities include the production of 'long,
structured, organised songs with themes, variations, repetition, rhyme,
and complex form.'

Some humpback whale song facts:

- The song has a predictable structure with a series of sounds
  (units), repeated over time in patterns (phrases), with each
  phrase repeated several times to comprise a 'theme'.

- A typical song is then made up of 5-7 themes that are usually
  repeated in a sequential order. A song typically lasts 8-15
  minutes (although it may range from 5-30 minutes), and
  then is repeated over and over in a song session that may last
  several hours.

- The sounds that comprise a humpback song are varied and can range from high-pitched squeaks to lower frequency roars and ratchets.

However, Rothenberg points out, 'other whales use sound in much less musical ways, orca, beluga, and dolphin vocalisations sound much more like a language. But humpback songs are long, very repetitive and organized utterances of sound. They are solo performances by males, much like bird songs which we tend to understand as musical performances that must be sung just the right way, over and over again.' David wonders why, 'like music, the songs do not convey much distinct information in the technical sense – if we've already heard the song why do we need to hear it again and again and again?'

Originally it was thought that these songs had a reproductive purpose and that they were simply a display of lek mating, which is where large groups of males gather to compete for attention, because the songs seemed to culminate when large groups of males would congregate. But time and more research have shown that the females don't have any involvement or response to the singing. Now whale researchers think it might serve the function of identification and association among males, so they can see who knows who, how they're involved and to what degree.

Thanks to the Cold War, we discovered the existence of marine mammal communication. Americans spying on Russian submarines, which were using low frequencies to communicate, picked up another kind of sound on the recording equipment. We now know that marine mammals use many and varied forms of vocalisations to communicate for much the same reasons that all creatures do including location, food, navigation, social bonding, reproduction and territoriality. While visual communication is limited for underwater creatures, they have the upper hand when it comes audio communication.

Even though it seems counterintuitive, sound can be heard across great distances in the ocean, in fact it travels five times faster than in air. Lower frequencies can he heard half way around the world.

It remains a mystery as to how baleen whales produce their low frequency thumps, moans, tones and pulses. While they do possess a larynx and vocal folds, it might not be the site of sound production, as is

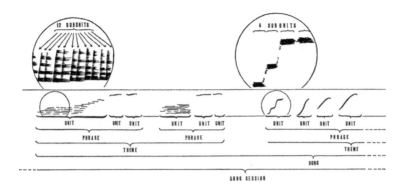

the case with dolphins. The sounds and methods of sound production vary from species to species. Dolphins and some whales create clicks, whistles and pulsed sounds from air sacs inside the head. The details of the sound production in toothed whales however, are a controversial topic. It is thought that air stimulates vibrations which is amplified by air sacs. These vibrations are then channelled through fat in the round organ in the forehead called a melon. The sound is then projected into the water and allows for very precise sound production.

Marine mammals use their unique methods of sound production for all kinds of reasons, often involving echolocation and mimicry. Mothers use long, beautiful songs to bond with their babies and male dolphins create unique calls when they leave home and find a best friend to swim with. The male humpback songs evolve over time, changing from month to month and year to year. They listen to the song of others, they repeat and modify depending on the situation. Even more perplexing is the fact that, in the same ocean, across thousands of kilometres, humpback whales will sing virtually the same song, as if they're all tuned in to the same hit parade of the deep.

The Whale Trust organisation's research strongly suggests that the songs function to facilitate social interactions between adult males, with no evidence of a female response to date. But Rothenberg isn't convinced. He explains, 'it may be about something else entirely ...

DIAGRAM: McVay

there must be some emotional reason we and the whales need it (music). Emotion is well-documented in all kinds of animals, all the way back to Darwin with *Expressions of Emotion in Animals*. We suspect birds are addicted to singing, because dopamine is released in their brains while they sing. Humpback brains also contain spindle neurons that had previously only been seen in higher-order primates, those believed to be able to experience complex emotions.'

Spindle neurons are associated with the ability to recognise, remember, reason, communicate, perceive, adapt to change, problem solve and understand. The latest research has shown that cetaceans have three times as many spindle neurons as humans. The intelligence of marine mammals is not disputed; they are well known for their complex social behaviour and abilities. Behavioural ecologist Peter Tyack believes that 'humpback song is a form of animal culture, just like music for humans would be.'

Back in the 1960s when we accidentally discovered the secrets of whale song, perhaps we weren't tuning in to the clumsy clicks, groans and wails of lumbering oceanic giants. Perhaps we were tuning in to a secret life, a sonic landscape that whales rely on to create meaning. Like folk songs in human history, whale song might serve important functions like establishing a sense of belonging, skill and position but it may also serve the purpose of enjoyment. If the ruling ideology of evolutionary theory is functionality, then why do they sing such long and complicated songs when something much simpler might suffice? Perhaps whales might also possess what we have wrongly assumed is an exclusively human desire for self-expression, creativity and perhaps even an appreciation of aesthetics.

———————————

Andy Coghlan, 'Whales boast the brain cells that make us human', New Scientist, November 2006, http://www.newscientist.com/article/dn10661-whales-boast-the-brain-cells-that-make-us-human.html#.U9DAIIBdV9Q

Peter Tyack TED Talk, The Intriguing Sound of Marine Mammals, Transcript online, https://www.ted.com/talks/peter_tyack_the_intriguing_sound_of_marine_mammals/transcript

http://www.whaletrust.org/

http://www.whoi.edu/

EXCURSION: RECORDING TECHNOLOGY

Expert insights on recording equipment and how to use it.

GEOFF SAMPLE: I came across the CDs of an American recordist – no names – and was disappointed to realise there was a certain amount of 'production' going on: looping of sequences, added reverb and such-like. Because I was trying to explore the subject matter, as well as enjoying the sonic textures, I really didn't want to hear it manipulated in the studio. And I was being drawn into listening to (and wanting to record) the full acoustic scene at chosen points of time and space. At the time, working in the studio on music projects, I was spending long periods programming digital reverbs to get something special on vocal performances, and I was knocked out by the subtle texture of the natural acoustics out there. And this has carried through into all my subsequent nature recording: I really value context in my environmental work, not just topographical acoustics, but involving the surrounding community. So often in natural and undisturbed habitats, there seem to be connections in how different creatures vocalise; and this leads into the nascent subject of soundscape ecology.

I was also drawn by the challenge of recording wild creatures as a live event, in contrast to all the over-dubbing that's become common-place in studio recording. You can't ask the wildlife to run it again.

I went for a Sennheiser mid-side stereo microphone set-up, which has about a 270 degree field of reception (i.e. a small dead area at the back); and I still use this. I like the discipline of single point stereo: it makes you listen carefully and think about the acoustic scene. You compose the soundfield in the exact placement of the mics and their direction.

I tend to set up the mics on a long cable, so that my presence doesn't influence the activity, or simply frighten off wary creatures. But I like to listen in, while I'm recording; and that's probably the most intense listening I do. Often I can't see the immediate vicinity of the mics, so it can be really immersive and exciting if you've got activity around the mics and you only have the acoustic cues to interpret what's going on.

PAUL PRATLEY, Secretary WSRS: I have met several birdwatchers over the years, who complain that the recording of a species is not true, because it was made too close or with a parabolic reflector, thus picking up detail of the song or call that the human ear is unable to distinguish at a distance.

I would say that we are recording what other birds would hear, and to understand their world we need to hear what they hear.

MIKE WEBSTER, Director Macaulay Library: The Macaulay Library started in 1929 when some people in the movie industry were experimenting with a new technology – adding sound to film (Talkies) – and they thought that filming birds singing would be a great test of that technology. Over the decades, the art and science of recording and studying the sounds of nature has continued to improve as technology has improved: the original film-based method of capturing sound eventually gave way to magnetic tape, which eventually gave way to digital recorders. The technology kept getting smaller and lighter and easier to use. Today it is very easy for somebody to take a walk in nature with some light-weight equipment and make excellent recordings. Indeed, most of the recordings in the Macaulay Library have come from enthusiastic (and talented!) bird watchers. At the same time, our ability to analyze these recordings has also improved dramatically. With computers we can now quickly scan many hours of recordings to pick out particular sounds. Research in bioacoustics is not reaching an end – if anything it is experiencing a renaissance brought on by the digital revolution.

JANA WINDEREN: I am using a minimum of two hydrophones, in this case DPA hydrophones (two to be able to hear where the sound is coming from). I use a Sound Devices 744T hard disc recorder. The sounds of the fish and crustaceans are relatively quiet compared to the sounds from for example a boat motor. So it is important to record when there are no motors close to the site. Often I have to record in the middle of the night when most people are sleeping to avoid the manmade sounds. So the sound level is relative. I have to gain quite a lot on the input of the recorder to capture these sound, so a good and quiet preamp is important. (I am for example now using a Reson hydrophone which is even more sensitive than the DPAs, so I can record the tiniest of underwater insects without getting the sound of the preamp.) Though the sounds from the decapods are amazingly loud compared to the size of the creatures.

When I work with the sounds afterwards I select those that are most

clear, and it is quite some work to tidy them up. Though when I am out there I move the hydrophones and change location to get closer to and capture the sounds that I hear are from a living creature. I am not saying I am not recording the sound of the current flowing through the sand and sea weed and other more 'mechanical' sounds. Concentrated listening is essential in the way I work. The composition process starts already out in the field.

I do not change the animal sounds themselves, I would not start to tune the animal sounds for example, I feel it would be disrespectful of the animal to use them as instruments. My point is not to make music from them, but to rather present them, although I am not trying to represent nature. I would much prefer to take the audience out there to the sounds to listen for themselves. Working with these sounds has made me a better listener, I think. It is also easier for me to play guitar now than it was before, I listen to the sounds, not thinking what chord I am triggering ... I think it also has given me a better ability to listen

PHOTO: (L) Yannick Dauby, (R) David Rothenberg

to a sound scape outside (or inside), to separate individual sounds, to hear movement in the sounds, to hear distance, and to hear spaces. I can almost always hear what kind of room someone is in when they talk to me on the phone, and recognise the exact locations for example ...

RODOLPHE ALEXIS: I had a four-track recorder, a surround system, a parabolic microphone and a hydrophone. This equipment was with me at all times. I followed a similar protocol for each site: First the ambiences with my surround setup: a four static cardioïd microphones system on a stand, head height, with a cross configuration which is in my opinion the best compromise regarding burden/mobility/impression of space. My installation and album *Sempervirent* is derived from the recordings carried out with this device. Then, after having placed the device on a stand, I would go hunting sounds with the parable, with the desire to find the best signal-to-noise ratio to be able to integrate the recorded animal in my installation 'Dry Wet Evergreen'. The rumble of the Pacific Ocean at Corcovado was a local problem for instance.

DAVID VELEZ: In the Palomino forest I found myself recording without headphones away from the microphones trying not to bother the animals with my presence. There I finally articulated emotionally and intellectually the importance that listening has for me in terms of the sensible experience. Like Bernie Krause said: 'A recorder is a tool for learning to listen without a recorder'. I experienced the poetry and meaning in the listening process away from my recording devices while joyfully and patiently waiting for the memory cards to be filled: I became aware that the resonance that one can establish with a certain environment is the key to the emotional success of all phonographic and in general to all documental-based sound art manifestations. My gear was a handy recorder, a tripod and as many batteries and memory cards as I could bring.

JEZ RILEY FRENCH: 'Micro-listening' is about really extending the listening range – focusing on & revealing the smallest sounds. For example, on the track I contributed to the *Favourite Sounds of Praha* CD I found that while listening to some very quiet recordings of spaces in the city (churches, parks) there were moments when small sounds associated with the time just after a microphone is placed or the record button is pressed caught my ear. In some instances these were sounds that some might edit out – the sounds of a microphone wind cover springing back after being in one's bag or indeed the bag settling when its used as a make-shift rest for a microphone – but I found the act of focusing in on them revealed such an intense sense of the ambience they were part of.

CHRIS WATSON: Do I ever worry about cutting up what may be a coherent 'song' to an animal into pieces that I may find appealing purely for aesthetic reasons? That's a good questions, and yes I do. The integrity of my work is of paramount importance to me. One of my tools for composition is editing and so I have to do that but I'm always very mindful of the integrity of the piece, in terms of habitat and species and the vocalisation. I'm aware of it and I try to resist the temptation to cut into stuff, but like with any project, the piece is not a perfect document in any way. It's a representation and that representation has my personal stamp on it, it has to have. I work and cut the piece in sympathy with the content but sometimes in terms of length, some things have to be sacrificed or constrained. It's like when you constrain the dynamic range of something so that you can actually hear it, or if you filter it. I would include all those devices with the term editing. Sometimes I've recorded animals that make infrasound which is sound below our frequency of hearing. For example, the glacier from the third track, 'Vatnajökull' on the *Weather Report* album contains infrasound and I had to edit that to bring out some of the content, pitch it up so that we could hear it.

What gets lost in recording is spatiality. That's the thing that we're just on the cusp of now. With ambisonics and surround sound we're able to record sounds that you can place into a three-dimensional environment. That loss of a sense of space is why I still go out and enjoy listening as well as recording.

There's so much information, beauty, interest and detail in the spatial aspects of sound. Animals use it all the time. They use it to define where they are in their habitat, and where their mates or rivals are in terms of distance, what the weather conditions are like and which way the wind is blowing, where the rain is coming from, the levels of audio from the stream beneath their feet or the wind up in the tree canopy. So it's definitely the way spatiality gets lost in recordings that I find most disconcerting. I'm continually trying to find ways of encapsulating that in what is still very much a two-dimensional stereophonic world. I like working with installations because we can install third order ambisonic systems for the performances. For this, I work closely with Professor Tony Myatt at the University of Surrey. We can bring spatiality into a performance space, which is really quite exciting for me.

I may edit animal sounds by cutting, splicing or filtering but I don't re-pitch things like that usually. With elephants you can make an exception to the rule, which is what I talk about in my workshops. They make infrasound, that has a fundamental frequency of around eight Hz, but it's a very complex sound. It's not a sign wave at eight Hz, and because of that it's harmonically very rich. So while we can't hear eight Hz, there's a harmonic at 16 Hz, but we also can't hear 16 Hz, but there's a harmonic at 32 Hz that we can hear, and the next one up at 64 Hz and there are lots of other frequencies around that. It's those complex harmonics that give a quality to the sound, whether it's a Stradivarius violin or a female elephant giving a 'let's go' rumble in the plains of Amboceli. We can't hear the infrasound but we can hear the fallout from it. It's true with insects as well, which have fundamental frequencies sometimes in the hundreds of kiloHz, that we can't hear. What we can hear is the harmonic fall out that gets down to 20 kiloHz and below.

SLAVEK KWI: Equipment employed: Sound Devices 722, Zoom H2, microphones DPA 4060, hydrophones Sensortech SQ26, Aquarian H2, Cetacean Research Technology C55 and C54XRS, ultrasonic range microphone Laar USM 20-6, heterodyne bat-detectors Laar TD05-B and Lars Pettersson D 200. It's not that easy to obtain good underwater-recordings. There is more to it than just throwing in the hydrophone and pressing record. Aside from the different acoustic properties of aquatic environment, temperature, salinity etc. there are plenty of additional sound sources to consider, all adding and changing recording conditions; for example: water (and canoe) is constantly moving, various debris are floating around and in the case of flooded rainforest, there are branches, uneven bottom and water also transports different sounds from distance and surface etc. It is rather challenging. As the echolocation of dolphins is ultrasonic, you cannot hear most of it – it happened to me that I recorded and didn't hear anything at all. I assumed there were no dolphins, only to discover later in the computer that I had picked up clicks of sonar.

EXPLORERS WITH MICROPHONES

*Who are the people and institutions that have dedicated their lives
to animal sounds?*

PLACES OF RESEARCH

The Macaulay Library and the Wildlife Sound Recording Society
are working on a deeper understanding of animal song and the
education of field recordists. Their ultimate goal: A comprehensive
dictionary of animal sounds.

If home really is where the heart is, then Mike Webster's home is his office: The Macaulay Library, the world's most expansive archive of animal audio- and video recordings. Located three and a half miles away from Cornell University's colourful hill-campus in Ithaca, New York and idyllically embedded into a rich, green wetland of trees, bushes, shrubberies and an over-sized pond, the library is part of the Cornell Lab of Ornithology's 90,000-square-foot Johnson Center, which furthermore hosts the Adelson Library and a state-of-the-art visitor centre. The sympathetically nested complex, mainly constructed from a lot of glass and stone, 'chiselled in place from fortuitously found materials' and accessible to guests by a wooden bridge, is almost like an island unto itself and is said to look like a bird from above. Being able to spend every working day in this inspiring environment is extremely gratifying. But as the director of the Macaulay Library, Webster also carries a huge responsibility. Not only does the library receive hundreds of thank you notes each year from amateur enthusiasts, who use its website to identify birds in their gardens. It has also become an indispensable source for researchers, with 600 scientific papers acknowledging the use of the library's recordings. As Webster points out, that number is growing exponentially each year, revealing a 'scientific impact that rivals any other biological collection.' It may seem as though Webster were living in a dream. But the implications of his actions are very real – and they're impacting the world of ornithology in previously unimaginable ways.

While visitors are spending their day looking out across the surroundings through the Morgan Observatory's two-storey-high window-wall or visiting the Johnson Center's interactive exhibits and theatres, Webster and his team are busily working away behind the scenes. Several staff members are processing licensing requests, which are coming through in large numbers. In a financially important development, approximately half of these requests are commercial, either from museums or developers of smart-phone apps and other software, publishers and even movie producers and other entertainers and artists – 'our sounds', Webster says, 'are often used in movies, television shows, and musical pieces'. The rest of the staff are responsible for the building and maintenance of the library itself, curating and archiving the audio and managing the rapidly growing database. Twelve years ago, the Macaulay Library also began including natural history videos and this new section of their collection has turned out to be extremely popular, today encompassing 50,000 entries from around the globe and many different species.

Seven generations of research

The central position of the Macaulay Library is unrivalled and its success is the result of seven generations of passionately dedicated researchers. Each director of the Macaulay Library – or Library of National Sounds (LNS) as it was first known – has set his own priorities and thereby both focused and expanded its scope: The legendary Arthur A 'Doc' Allen was the first to champion the use of sound recordings, rather than paintings, for studying avian behaviour. Albert R Brand, together with Peter Keane, developed the first portable (albeit still extremely heavy) field recording gear and produced *American Bird Songs*, a popular collection of recordings. Later, Peter Paul Kellogg and Doc Allen would host their own radio show about ornithology, *Know Your Birds*, which would play a significant role in popularising their cause. James L Gulledge recognised the vital importance of funding and sponsoring, the fruits of which his successor Robert Grotke eagerly made use of by expanding the archive to a staggering 140,000 recordings.

Since Grotke's expansion, the key tasks have consisted in making full use of the digital domain by digitising and tagging the material and continually improving its accessibility and searchability. Its three pillars can be summarised as 'Education', 'Outreach' and 'Research', expressing the constant connection between the popular and the scientific, between collecting data and making it available, between practical actions in the present and furthering environmental topics in the young. To this end, the Macaulay Library has consciously moved away from its ornithological specialisation, as Webster stresses, and today features recordings of 10,000 different species, of mammals, fish, frogs and insects.

And yet, one goal has kept this diverse mission together, as Webster points out: 'Many animals communicate with acoustic signals, and by studying those signals we learn a great deal about how evolution has shaped them, about how animals produce the sounds that they do, and about the things that animals say to each other. We want to understand all of this – how and why animals communicate with each other – but to understand their language you need a dictionary. The Macaulay Library is one such dictionary.'

In fact, it is rather like a dictionary, encyclopaedia and chronicle all in one, allowing for long-term observations on an international scale: 'We can learn how the environment itself affects the ways that animals talk to each other. Today, animals communicate with each other in a world that is increasingly influenced by human activities, including the sounds that we make. The noise pollution that we generate reaches virtually every corner of the globe, and animals must deal with it. By studying recordings in the Macaulay Library we can learn *how* they are doing so: we can compare recordings made decades ago to those being recorded today. We can learn how birds change their songs if the environment becomes increasingly urbanized, how whales deal with the sounds of ships, and how elephants modify their behavior when humans move into the area.' To Webster, therefore, the dimensions of the library's work are far more universal than generally acknowledged in the media. It may be surprising for some to hear him claim that singing is never just a purely functional process for birds, but always about joy and aesthetics as well. Those that claim otherwise, however, aren't necessarily taking a more rational stance. If Webster claims that there is an emotional and genetically implanted appreciation of sound and this is what connects humans and other mammals, then making us more aware of these ties means changing our attitude towards our environment. And if our attitude is what determines our actions, then Webster's work isn't just scientific – it is always political as well.

Medium-sized community

The same can be said, albeit on a smaller scale, about the Wildlife Sound Recording Society. Founded in the UK, the organisation today boasts members from every single continent bar Africa and the polar regions as well as an important affiliation with the British Library, which has been a great help in spreading the society's work far beyond the niche of field recording enthusiasts. Only 300 members' strong, the WSRS is nonetheless a great example for one of many medium-sized communities contributing to the research of animal behaviour, -sounds and -communication. It has placed great importance on displaying professionalism, publishing a semi-annual magazine and four collections of recent recordings per year, next to organising meetings and topical debates on their core topics. In 2011, it released a double-CD set called *Wild World* – a global overview of wild places, containing 22 contributions from its members.

In many ways, the album represented a sonic summary of the society's work and demonstrated how international its reach had become: Featured on the release, among others, were birds from the Vwaza Marsh in Malawi as well as the Analamazaotra park in Madagascar, from Mount Lewis in Queensland, Australia and the Twin Lakes near New Liskeard, Ontario, Canada, from Hungary's Bükk Hills to Poland's Lake Zawadówka. And, of course, it also included various snapshots from England, the WSRS's home – an important aspect, since, as the society's secretary Paul Pratley, explains 'most of our members are fascinated by the everyday sounds around them.' Pratley's own road towards membership may be typical in this regard: 'My very first experience of wildlife sound recording was when I was a teenager. My father took me to the local bird group's open day. There for the first time I was given the opportunity to listen to some recordings using headphones. From that day onwards I wanted to make recordings like that', he says, only to dryly admit: 'I have yet to think I have achieved it.'

At the same time, his ongoing quest for improvement is what motivates the society's members, next to providing a platform for like-minded bird lovers and recording fanatics. The WSRS officially sums up its goals as 'encouraging the recording of wildlife sounds', 'offering opportunities for technical improvement and scientific study' and 'furthering the appreciation and understanding of animal language'. In practise, meanwhile, its impact is felt in three different, albeit closely associated areas. For one, its community provides for an informal and extremely efficient transfer of expertise and knowledge. Secondly, it has, since its inception in 1968, collected a wealth of data about the recording aspect of sound collection, gathering valuable insights on microphones, portable recorders, recording techniques and editing technology. And finally, it has established quality standards in animal recordings, which have been extremely important in its scientific reception.

Recordings are sent to the British Library for assessment and analysis and for making them accessible to the public, for example. It has also raised awareness for non-interfering recording approaches: While pishing – the careful human imitation of bird calls to lure them from their hiding spots – is deemed acceptable, calldown, the playback of recorded sounds in the field, is discouraged, as it causes distress and distorts natural sound. Pratley stresses the vital importance of keeping to certain formal documentation requirements, not just for scientific usability, but also in terms of their informative value in general: 'There are many sounds from nature that are quite audible if you get close enough to the subject. One that springs to mind was of a recording of an adult curlew *Numenius arquata* brooding its eggs, the eggs were on the verge of hatching. The sound of the chick could be heard inside the egg, also the very soft calls of the adult responding. There would be no other way of hearing these sounds unless you had a mic close to the subject. Any recording made no matter what it is, if it is not documented, is meaningless, no matter how good it is.'

Contrary to Mike Webster, Pratley doesn't believe in the concept of music or the notion of aesthetics outside of the human realm. And yet, hearing him speak about his most memorable and favourite bird songs does reveal his deep emotional – as Webster would say, evolutionarily forged – connection to these sounds: 'I spent many years living in Norfolk on the East coast of England. There are two sounds that take me back there; one is the bubbling call of the curlew out on the marshes at night. The other is the call of pink footed geese. Skeins used to fly over the offices where I worked on winter evenings to their roosts in the Wash. Ten thousand pink feet flying in the half light, just over the tree tops on a winter's evening with their plaintive honks is a most haunting sound, and something I will never forget.' Thanks to the activities of the Wildlife Sound Recording Society, these memories, and the emotional potential contained within them, are now open to the public.

---

http://macaulaylibrary.org/history

http://www.wildlife-sound.org/about/index.html

FATHER FROG: YANNICK DAUBY

Frogs want to be heard – and Yannick Dauby loves to listen to them.

Night has fallen over Guandu Nature Park in Taipei. Yannick Dauby is quietly striding through one section of its 57 hectares of ponds, wetlands and brackwaters, carrying bags brimful with equipment and a boompole. For Dauby, this is the best time of the day, the moment human noises slowly ebb away and the heat becomes somewhat bearable, the moment the snakes come out. The latter, as the French sound artist will later tell me with palpable enthusiasm, are 'part of the pleasure' for him. But Dauby isn't looking for snakes, at least not tonight. He is driven by the memory of an encounter with an outwardly far less impressive creature: *Rhacophorus moltrechti* is so small that the uninitiated observer probably wouldn't even notice it unless you were standing right next to it – it's roughly the size of a child's pinkie. With its garish-green skin and reddish pants, many might just as easily confuse it for its cousin, *Hyla arborea*, the European tree frog. *Rhacophorus moltrechti* isn't even present in Guandu Nature Park. And yet, to Dauby, it has taken on an almost spiritual significance.

The story of his first encounter with *Rhacophorus moltrechti*'s hard-to-resist 'soft timbre, round-shaped and woody character and its three descending tones' dates back to an August day in 2005 in the Southern mountains. Even when he's speaking about it now, it sounds as though he's describing some kind of personal epiphany, the kind of life-changing experience that some might associate with marriage or the birth of their child: 'At noon, while the cicadas (*Cryptotympana takasogona*) were singing loudly, I heard a surprising sound four or five times. I was surprised and recorded a very short sample before it stopped. I came back to France shortly afterwards and had to wait one year, before Professor Yang Yi-Ru introduced me to this fascinating endemic tree frog. It became a kind of personal mythological sound that perfectly represents my experience of Taiwan.' Since then, Dauby has spent countless hours recording the sounds of the frogs of Taiwan. Eventually, the results of his endeavours were published on a now sold-out CD containing 16 pristinely realised audio portraits. That said, the project is far from completed.

Back in Guandu, after wandering about seemingly aimlessly for a while, Dauby has found a nice spot and starts unpacking his gear in the moonlight. Although it has by now become significantly cooler, special condenser microphones, capable of resisting the humidity, are still required. So is the boompole. Frogs may be easy to record once you've found them, being sensitive to movement rather than shapes and colours ('Just stay still, and they'll restart singing', according to Dauby), you still need to get as close to them as possible. Here, the boompole comes in handy: 'There is nothing like the feeling of having a microphone located a few centimetres next to the mouth of a nice frog! It's like putting my ear just above the strings of a musical instrument ...' Needless to say, not everyone shares his excitement. The sound of even some of the more common species can reach ear-spitting levels and since frogs have the tendency to reach their peak at night, in Germany, for example, there have been numerous lawsuits initiated by sleepless frogophobes trying to force their neighbours to fill up their ponds with concrete and chase the animals away. Of course, biologically speaking, frogs have to be this insistent to stand a chance of reproducing. Since the calls of male frogs are easily drowned out by the myriads of competitors as well as the noises produced by other species, mating has turned into a survival of the loudest. Nonetheless, the disparity between their unassuming size and volumes of more than 75 decibels, remains remarkable. And because of fascinating facts like this, Dauby still hasn't lost his pure love for spending time near the frogs without any conceptual considerations. And as he slowly moves his boompole towards a frog, that's precisely what he's going to do tonight.

A second home

In terms of recording, tonight's excursion is an exception. With large roads nearby creating a constant rumble, Guandu Nature Park, despite its undeniable beauty, is more of a meeting place and home base for a small but tightly-knit community of animal scientists, ornithologists, herpetologists or entomologists. And so, to capture the

PHOTO: Wan-Shuen Tsai

best frog sounds, Dauby has journeyed all over Taiwan: 'I recorded *Rhacophorus* in the Alishan mountains, the beautiful *Rhacophorus auriventris* near Taitung in the South, the rare *Babina okinawa* in a tiny pond in central Nantou, the common toads (*Duttaphrynus melanostictus*) in Peng-Hu archipelago and even in the campus of the National Taiwan University of the Arts in Taipei! Actually, I often go for walks with the other volunteers of Guandu Nature Park in the surrounding hills of Taipei, which are full of amphibians because of their high humidity. I also had the immense chance to spend some nights in Fushan Botanical Garden, which is a wonderful nature reserve.' And yet, in terms of conducting his research, there really couldn't be a better place for Dauby than Guandu. In a way, it has turned into a second home for him. Not only is he spending a lot of time in it as a field recordist. But since he moved to Taiwan in 2004, he

ANIMAL MUSIC

has also worked here as a volunteer. There is a lot of work to be done: Located in the North of the country at the junction of the Danshui and Jilong rivers, environmentalists fought for decades to convince authorities of turning it into a natural preserve and now they've been given their chance, they don't intend to waste it. An important first step was taken in 1983, with the creation of the Guandu Waterbird Refuge, although economic aspects of tourism still played into the decision. In 2001, however, a major breakthrough occurred when management of the park was handed over to a Non-Governmental Organisation, the Wild Bird Society of Taipei. With all proceeds flowing back into the park, it has since flourished and turned into a national symbol of pride. With 229 different species at present, birds are still unsurprisingly the most important animal here. But it is precisely the astounding all-around biodiversity that makes the enclosure so valuable. 'Wetlands are important because they are the habitat of a large number of animal and plant species'. According to Guandu Nature Park administrator Lin Yi-tzai, 'The volume of life wetlands can support is 2.5 to 4 times that of a regular piece of land. If they are well preserved, the biodiversity that they host would also be preserved and the soundness of the world's ecosystem can be maintained'.

It is a diversity visitors can not just see, but hear as well. Especially for Dauby, who grew up in a dry and arid Southern region of France, with only a single frog species singing on a regular basis, the wealth of sounds to be witnessed is almost intoxicating: *Hylarana latouchii,* with its low volume growlings and occasional click sounds. *Microhyla fissipes,* whose percussive song is excruciatingly loud thanks to a vocal bag the size of its entire body. *Kurixalus eiffingeri,* which produces a series of soft notes separated by a few seconds of silence. And then there is the particularly fascinating *Hylarana swinohei,* a 'quite big frog living in rivers and streams where the water is active, and therefore full of the oxygen required for its reproduction. Their call is well adapted to these noisy environments because it is short, intense and very high-pitched'. *Hylarana swinohei* constitutes an intriguing case of frog sound adapting to their environment. Other than that, the encounter with civilisation hasn't always been equally positive: 'I read at least one paper about frog communication changing over time due to human contact. Human-made noise affects the acoustic behaviours of some amphibians, and therefore decreases their chance for reproduction. For sure, frogs are very sensitive to human presence, even if sometimes they seem to ignore us. They feel vibrations from the floor, which is why they stop singing when we're getting close. But much more importantly, they are extremely sensitive to the quality of the water. Pollution and reduction of aquatic areas are the main threat for amphibian reproduction, noise pollution is definitely a lesser danger to them. Which doesn't mean we should ignore it.'

The added benefit of working at Guandu Nature Park is that it has allowed Dauby to work with like-minded individuals who, just like him, intend to make sure no one ignores it. Together with local environmentalists, he takes regular strolls to specific sites to monitor the presence and development of particular species. Since listening is a fundamental requirement for detecting the animals, his field recording work is greatly appreciated by officials. And yet, his hearing pales compared to that of one of Taiwan's leading specialists in the field, the very same Professor Yang Yi-Ru who revealed the identity of his favourite frog, *Rhacophorus moltrechti*, to Dauby. In fact, her expertise on frogs is by some described in all but legendary terminology and has earned her the title 'Mother Frog' at home. Stories about her have the air of fairytales: Once, Yang discovered 19 different frog species around a single pond – more than half the entire known amount in Taiwan. And once, as the *Tzu Chi English Quarterly* claimed, she even performed CPR on a frog: 'Yang once came upon a dying frog near a spa. Without hesitating for even a second, she pressed its heart with her finger and breathed into its mouth. After a few moments of this CPR, the frog miraculously came back to life. It's hard to question Yang's love for frogs after hearing this story.'

To Dauby, meanwhile, it wasn't so much these anecdotes that kindled his admiration, but the professor's uncanny ability of drawing a wealth of information from frog sound, a knowledge impossible to acquire from theoretical studies but merely from hours and hours of sitting in the field and listening. It is a kind of knowledge, too, which transcends the borders between the conventional term 'music' and undiluted nature sounds. Embedded into these signals, which can take on a bewildering beauty at times, are encoded messages, which can be exciting to decipher. It is all the more intriguing, since the accumulation of individual voices does not result in a cacophony, but rather a fascinatingly unified sonic panorama. The organisation of a chorus of frogs has to do with the superimposition of different cycles, creating an effect similar to phase desynchronisation, explains Dauby, who enjoys how a complex sonic architecture can result from seemingly primitive acoustic elements – an organisational principle he recognises from the world of contemporary composition. And then there is the sensation of being placed inside a mesmerising acoustic space. Since amphibians can't fly or run when singing, their chant represents a kind of static system, where a sound comes from precise position and creates a fascinating sensation of geometry. It may appear hard to detect individual voices amidst the chorus. But in fact, as Dauby points out, some species, including tree frogs, will often choose a spot from where it'll be easy to be heard. He puts on his headphones and flashes a smile of anticipation. Then, he presses record.

Guandu park to organize wetland conservation talks by Meggie Lu http://www.taipeitimes.com/News/taiwan/archives/2009/05/03/2003442653

Fan Yu-wen, Yang Yi-ru 'Mother Frog' http://enquarterly.tzuchiculture.org.tw/tzquart/99spring/qp99-17.htm

History of Guandu Nature Park http://gd-park.org.tw/en/history

http://www.kalerne.net/yannickdauby/

IN LOVE WITH THE OCEAN: HEIKE VESTER

Despite her fascination for seismic noise and the sounds of whales,
Heike Vester still foremost considers herself a pragmatic biologist.

Every year between February and April great colonies of sea lions gather to breed along the beach at the Valdes Peninsula on the North East coast of Argentina. Since 1976 a group of orca have been observed gathering offshore at the same time, their towering black dorsal fins parting the sea as they watch and wait. These animals have perfected a unique hunting technique which involves catching a wave into shore, intentionally beaching themselves as they snap up a seal pup playing in the shallow waters, before waiting for a new wave to help them manoeuvre back out to sea. Once safely back in deep water, the orca may share the pup, they may also play with it using their tails to slap the tiny carcass many metres into the air.

How did these animals learn this sophisticated technique? How do they know it is time to gather off the coast of Argentina? And how are they able to teach their young this complex and risky hunting strategy? Are they playing a game with the seal carcass? Or using it to teach the younger whales some sort of lesson? Heike Vester is trying to unlock the mysterious habits of these highly intelligent animals by observing how they live together and recording their communications.

Born in Germany, Vester relocated to Norway in 1999 to be closer to important whale sites. She said while the ocean hooked her at the age of 13, her love affair with the sounds of the ocean began in 1998 when she met her first bio-acousticians. 'They were studying sperm whales. They communicate only in clicks, but it was enough to catch my interest!'

The following year Vester moved to Northern Norway and studied killer whales during the winter months from 2003 until 2008, becoming obsessed with trying to understand how the animals communicated.

'[Killer whales] live in complex social groups all their lives (most in matrilines), they grow old, up to 90 years, they are capable of vocal learning, mimicry and pass information on to others and the next generation,' she explained. But despite decades of study scientists are little closer to understanding the beeps and sweeping calls of these animals. 'They use complex vocal communication, which even after 40 years of research nobody really understands, it is a challenge and great fascination to investigate their language,' Vester said. In 2009 the 'best of' Vester's marine recordings were released by Gruenrekorder as *Marine Mammals and Fish of Lofoten and Vesterålen.* Coming in at almost an hour long the release, which Vester says she does not classify as 'music', begins with the sounds of orca carousel feeding on herring, before introducing us to pilot whales calling, sperm whales echo-locating and a seal pup crying for its mother before closing with a track documenting that new type of whale hunt, whale watching.

Groans and tics, ascending and descending squeaks, electric pulses and the faint sound of a killer whale tail striking out to stun the fish it is hunting proliferate throughout the recording, painting a portrait of a profoundly different type of world, a dark place where sound is used not just to see, but to teach, to play and to survive. But as Vester underlines in the final track, it is a world which also bears the traces of human influence. Not just the tangible realities of pollution or litter, but the less obvious incursions such as the persistent presence of whale watching boats, or even more subtle interferences into the fabric of this dark place: noise.

The liner notes for the final track contain a plea: 'When we produce noise with motorized boats, we pollute the whales' environment and make it difficult for them to communicate. This may interrupt their feeding and social behaviour. We must continue to be vigilant, always consider our reasons for approaching whales, and most important of all: How?'

Although it has been receiving more attention in the media recently, ocean noise and its impact on whale communication in particular is a very poorly understood area. Military sonar is believed to have a particularly tragic effect, having long been linked to whale strandings. The PETA website, for example, states: 'The navy's testing of sonar systems has been linked to the beaching and deaths of marine mammals,' adding that following sonar tests in North Carolina in 2005, 39 whales perished. The arrival of 200 melon-headed whales into dangerously shallow Hawaiian waters coincided with a US-Japanese training exercise in 2004. 'The sonar is believed to disrupt marine mammal communication and navigation systems,' PETA states. But it isn't just sonar which can impact on whales, other more subtle noises can create less than ideal conditions for these giant creatures. Says Vester of her recordings: 'Sometimes you hear the boat before you see it. Seismic noise travels very far, I recorded seismic noise 300-400 km away! People, including myself, usually do not know about the impact of noise in the sea!'

But we also do not know much about the role of the noises of the sea which are meant to be there, and have been part of the sonic landscape of the deep for perhaps millions of years.

The chirping of dolphins, the grunts and snorts of fish, the startling electrical squeals of the killer whales, what are they saying? Do they ever just playfully communicate? Perhaps they even sing?

Much has been written about male humpback whales to all appearances hanging upside down in the ocean and 'singing' to attract a mate. Could these animals be making music? It is an amazing thought.

The more scientific among us, like Vester, don't seem to want to speculate about such things. 'I am more like a pragmatic biologist, the sounds are there and how we hear it and what it triggers in us or other animals is really up to the receiver,' she said, when asked if perhaps killer whales used sounds to make songs.

The sounds the killer whales make after eating certainly seem mellower and more contented even to these human ears than the shrill excited screeches. An after dinner song perhaps? Maybe we will never know, but these whales, like the ocean that they live in, certainly offer us a wonderful opportunity to dream and envisage the ways we, as land bound creatures, are so different from these animals of the seas. And yet, perhaps, also in some ways the same.

HACKING INTO THE SHRIMP NETWORK: JANA WINDEREN

In her quest of understanding the mysterious messages of crustaceans, Jana Winderen has set the scientific world on fire.

The world of shrimp professors is rarely one of excitement, breaking stories or emergencies. And yet, one day early in 2008, their Norwegian network was suddenly teeming with activity, phones ringing all across the country to the sound of heated debates and confusion. The sudden outburst of energy had been caused only a few hours earlier by a seemingly innocent phone call by a sound artist by the name of Jana Winderen. Winderen had been out trying to record the sounds of a huge school of herring with her colleague Chris Watson to the North of Bergen as part of their collaborative project *Voices from the Deep*. Two of the four days spent in a small boat borrowed from the Institute of Biology's Arne Johannesen were unsuccessful, since the gushing winds and the water lapping onto their boat were proving too loud for them to still be able to capture the sounds of fish. But the remaining time had proven to be highly fruitful and they had returned home with several hours of high-quality recordings.

Still, one thing had bothered them. Already on their first field trip, they had heard something that sounded suspiciously like snapping shrimp, widely considered to be one of the main noise sources in the world's oceans. Although Winderen was familiar with its characteristic sounds and had often encountered it on previous occasions, they weren't quite sure this time. And then, there was another problem with their observation: Typically preferring tropical and temperate waters, the coast of Bergen clearly seemed to be too far North for these crustaceans. Which left the artists with a nagging question: If snapping shrimps weren't causing these fascinating sounds – what was? For days, the shrimp professor network sent out a request to answer her question, evaluating and re-evaluating existing data, interpreting and re-interpreting what they knew or thought they knew and considering and discarding a plethora of hypotheses. Finally, they got back to Winderen back with their conclusion: They had no idea.

A sense of mystery

To Winderen, this open acknowledgement only added to the mystery she had always felt in relation to the sea. When interviewing her with Ed Bendorf about her installation and studio album *Energy Field* in 2010, I could sense the unbridled excitement for her work, her mission, her dreams. She was now, she told us, conducting vertical recordings with three to four different hydrophones at a time and looking forward to using a new, ninety-metre long cable to enter as yet unexplored sonic depths. She talked about how male cod were capable of producing sounds as low as 20 Hz, at the very limit of the human hearing range. She was fascinated by the fact that one could hear someone walking on an ice-coated lake up to 25 metres underneath its surface. And, while in Greenland, she happily placed the hydrophones into the ice for several hours to listen to it quietly speaking to her, only to discover a sign saying 'Do not enter, extreme danger to life' on her way back. Yes, she did want to make listeners aware of ecological issues. But rather than stun them with hard facts about the destruction of these complex and fragile worlds, she intended to engage them emotionally – just the way she was emotionally taken by the sounds of the water and its inhabitants as a little girl growing up by the sea in the South of Norway. In many respects, this dual aim explains her outwardly paradoxical decision to follow up her studies of mathematics, chemistry, biochemistry and fish ecology at the University of Oslo's Natural Science Department with courses in fine arts and design in England and eventually establishing herself as a sound artist. As such, in her creative capacities, she is capable of combining the scientific and aesthetic and arriving at far more captivating statements than a biologist ever could.

These considerations notwithstanding, it would take until her first encounter with cod in 2006 in Hardangerfjord, as part of a project for *Grieg07*, that hydrophone recordings first started openly appearing in her work. Maximally refined and delicate and somewhat inconspicuous on a first listen, *Surface Runoff*, a barely eight-minute-short 7 inch single published in 2008 on Touch, paved the way for a far more attention-grabbing and demonstratively titled tape a mere year later. *The Noisiest Guys on the Planet* (Ash International) was realised in the wake of her aforementioned trip with Watson and on board the Johan Hjort, a vessel of the Institute of Marine Research equipped with 'advanced acoustic instruments for fish detection and echo integration'. Her main ambition on the expedition was not to gather pristine-sounding field recordings, but simply to learn: 'The experience of seeing how systematically the fishermen and the marine biologists worked and how they were conducting the systematic fishing on each "station", every two hours throughout the day and night was extremely inspiring', according to Winderen, 'as was the experience of the challenges which appear when you are out in pitch black darkness in huge waves, sending out the trawler and the deck of the ship and everything around you is covered with ice.'

None of these memories is present on the tape, which, consisting of two 19-minute collages of decapod recordings, takes listeners into an artful – since all sound sources were carefully arranged in the studio – yet visceral and highly realistic sonic representation of what things are like underneath the surface, where to-us-alien-looking creatures are producing a living, planet-spanning installation of rhythmical crackles, snaps, pops and stridulations. Although a review at the time described the result as possibly the creepiest sound imaginable to human ears, the results were, in fact, rather intensely intimate and bizarrely comforting – Winderen herself simply refers to them as 'beautiful'. And yet, to her, *The Noisiest Guys on the Planet* is more than just sensory stimulation. It is an open call for action to explore, investigate and, ultimately, understand.

Hearing after all

Winderen isn't alone in her surprise at the gaping void in our insight into the acoustic qualities of the sub aquatic world. When, in 2011, an international team, put together by the University of Bristol, found that crustaceans could detect and avoid reef noise, it took the scientific community by surprise. For decades, it had been silently accepted that, with the exception of some types of crabs, decapods were deaf. Two years prior to the Bristol paper, Jennifer Taylor and Sheila Patek had already summed up the status quo in an unapologetically open statement: The 'mismatch between knowledge of sound production mechanisms and reception mechanisms is puzzling', they admitted, 'and can probably be ascribed to biologists not listening for substrate-borne sounds, although they are both present and probably widespread in crustaceans.' In simpler terms, their conclusion amounted to this: Scientists were looking for ears in an environment where ears could hardly be considered the most suitable organs. Instead, crustaceans had developed their own mechanisms for hearing, which differed considerably from those of humans and other mammals.

In an article on their webpage, the Museum of Victoria has described these in clear terms: 'When we hear a noise we are essentially listening to sound waves, which are changes in air pressure. Crustaceans do a similar thing, but they "hear" changes in water pressure around them. Instead of ears they have tiny (microscopic) hairs all over their hard shell, and particularly on their antennae. There are several different types of hairs, each connected to a nerve connected to the nervous system. One type of hair responds to physical stimuli such as water movement, vibration or touch. These hairs are called mechanoreceptors and they respond to a stimulus by sending a message to the nervous system. Some of the mechanoreceptors respond particularly to vibration or changes in water pressure – these are the "hearing hairs". A different type of hair is sensitive to chemicals in the water, and these are the "smelling hairs". But even with these hearing hairs, crustaceans hear differently from humans. If you play soft music to a crab, for example, the crab might or might not change its behaviour. But if you jumped up and down near a crab it would be much more likely to respond – usually by trying to escape from the source of the vibration. So crustaceans are often more responsive to vibration rather than noise – and that is how they hear.'

If the precise mechanisms behind crustacean hearing still remain poorly understood at best, then this is partly because scientists are finding it hard to decide where to start. With the breathtaking diversity of different species, the recent discoveries have opened a Pandora's box. Some sound-production mechanisms are shared by various species and allow for categorisation. Others, on the other hand, appear unique. The snapping shrimp, for example, closes its claws so viciously and fast that the water around it cavitates, creating a tiny vapour-filled bubble.

Its collapse produces not just a sharp sound, but also a tiny spark of light, invisible to the naked eye – a phenomenon known to intimates as 'shrimpoluminescence'. As Bernie Krause puts it: 'We can compare the shrimp output to a symphony orchestra, which may generate loud moments peaking around 110 dBA. Indeed, the lowly, unsophisticated shrimp will not be outdone even by the Grateful Dead, a rock group whose concerts have been measured at levels exceeding 130 dB. Get this, Deadheads: the shrimp is louder by close to a factor of five – all that without a huge stack of stage speakers!' Spiny- and clawless lobsters, meanwhile, as Winderen points out, emit sound by means of their antennae in stick-and-slip motion, similar to the way a violin bow moves over the strings of the instrument. This form of communication isn't just shaped by its sonic properties, meanwhile, but by evolutionary considerations as well, making research on the topic even more difficult: 'This methods make them less vulnerable when in their soft shell phase', Winderen explains, 'since they are not depending on the hard shell to make the sound, as are other crustaceans when they make the sounds by rubbing two body parts together.' The reasons behind the mysterious sounds penetrating *The Noisiest Guys on the Planet* and *Energy Field* remain even more opaque. Some of them may merely be feeding noises, some may signal proximity to other members of the group. Others remain a complete enigma.

Winderen has her suspicions about this lack of insight, attributing it to a lack of commercial interest for the fishing industry – it is telling that up until recently, it was merely the military who would conduct tentative explorations into the subject matter, since the sounds produced by the decapods are capable of covering up submarine noises. As ecological topics are becoming more and more political by the day, this may be about to change. Once, Winderen was called by a German journalist who wanted to corroborate that the snapping shrimp had indeed, as he'd been informed, moved further North in response to climate change. There was something strangely satisfying about the call. A beginning had been made, heralding more inquiries, more interest and more awareness. It may have been given some rest for now – but the shrimp-professor network may have to get used to increased attention very soon.

Interview with Jana Winderen by Ed Bendorf & Tobias Fischer http://www.tokafi.com/15questions/interview-jana-winderen/

JRA Taylor and SN Patek. Crustacean seismic communication: heard but not present? In: The Use of Vibrations in Communication: Properties, Mechanisms and Function Across Taxa, ed. C. E. O'Connell-Rodwell, Research Signpost, 2009

University of Bristol. 'Rowdy residents warn crustaceans away from perilous reefs.' ScienceDaily. ScienceDaily, 10 February 2011. http://www.sciencedaily.com/releases/2011/02/110207112816.htm

Can crabs hear? http://museumvictoria.com.au/discoverycentre/infosheets/can-crabs-hear

Bernie Krause: The Great Animal Orchestra

THE MECHANICS OF MUSIC

How do animals produce sounds and how do they sing?

ON THE VERGE OF DELIRIUM: BIRD SONG

We still don't know how, exactly, birds produce their songs and
what they mean. Perhaps our human brains are simply not
up to the task.

When it comes to bird song, the rules of attraction are always the same: One sound and you're hooked for life. In the 1980s, Geoff Sample had moved from his parents' home in the countryside to London, where he began working in a small recording studio, producing mainly for solo artists. Longing for the soothing presence of nature, he tried sneaking out from his busy schedule for short excursions into the Scottish Highlands. It was here that he first started listening to bird song in the early mornings and that their calls took on a new meaning for him. Suddenly he was struck with the momentous realisation that the warblers he was hearing weren't just mechanically repeating lines, but constantly varying them. The discovery knocked him out with excitement and he had to find out more. Birds would turn into a life-long passion for Sample. He began his journey into the subject at the best possible moment. Portable DAT recorders had just become available at reasonable prices and they allowed him to commit the songs he was hearing to tape in order to verify his hypothesis: Was he just imagining it, or were they really improvising?

Heroic pioneers

As Sample would quickly discover, he was neither the only one embarking on a mission, nor the first. Little wonder: understanding sounds is vital for understanding wildlife in general and is therefore of seminal importance for anyone spending a lot of time in natural habitats. As Paul Prately of the Wildlife Sound Recording Society points out, sound can signal the presence of predators; a high-pitched seep from a blackbird, for example, may mean that a sparrowhawk is flying overhead. With no species, however, are sounds as important as with birds. Typically, they are even identified by scientists not by their appearance, but their sonic repertoire – partly for practical reasons (many smaller bird species are all but impossible to sight), partly for biological reasons (dialects and specific vocabularies can indicate genetic differences where one wouldn't see them). And so, over the years, many generations of curious ornithologists have ventured into the field to expand their knowledge and understanding and to document the birds' acoustic world as precisely as possible. Especially in the early years, when equipment was less portable than it is today and the trade was still in its infancy, their efforts were truly heroic. Which may explain the legendary status some of these pioneers like Jean Roché and Walter Tilgner, who were among the first European field recordists, command today. To both, birds weren't the only points of interest. Roché started out filming insects with his camera at the age of 22, while Tilgner has always been fascinated by the forest as an acoustic body. However, for almost 50 years, they have repeatedly returned to bird song, documenting their explorations in vast discographies ranging from mostly local collections (Tilgner) to South American and Asian catalogues (Roché).

Yet their star was easily eclipsed by a man reverently referred to either as 'the Mozart of ornithologists' or simply as 'by far the greatest specialist on the life histories of neotropical birds there ever was': Theodore 'Ted' A Parker III. To this day, Parker has retained a semi-mythological aura. According to Russell Mittermeier, chairman of Conservation International (CI) in Washington, DC, Parker and his 'rapid assessment' team 'carried two-thirds of the unpublished knowledge of Neotropical biodiversity in their minds.' This knowledge was the result of an obsessive drive to gather as many bird songs as possible. In 1971 alone, he recorded more than 600 species. Parker possessed a supernatural memory for sound and place – he could reportedly identify a recording as specifically as coming from the 'south bank of the Amazon between the Rios Madeira and Tapajos' – and an ability of combining state-of-the-art recording technology with traditional techniques; described as 'walking slowly down a path, frequently pausing to watch and listen, picking out single calls from the flood of bird song'. Bird song was Parker's life and he invested every single second available to him in his passion. By the time of his death in a plane crash in 1993, he had captured the sounds of more than 1,600 species of birds and made over 15,000 recordings.

The work of these early researchers remains vital. Sample, for example, found essential confirmation of his own work in the recordings of Tilgner: 'He delivered a real sense of being there. He records binaurally with a Neumann dummy head and publishes his recordings with minimal processing or editing. His choice of subject matter was also revelatory for me.' And yet, one can't help but notice that general interest in bird song has been waning for decades. Sample theorises that the decline in the caged bird trade is not so much the result of animal activism, but rather of radio and recorded music replacing birds in their function of providing pleasant aural wallpaper. To the layman, too, most of the scientific insights gathered over the years have remained elusive. It is telling in this regard that one of the most popular books on bird song, Alwin Voigt's *Exkursionsbuch zum Studium der Vogelstimmen*, was first published in 1894 and has since been re-printed again and again with only minor additions. Bird song has become the domain of older people, and interest in it has occasionally taken on slightly bizarre traits: 'I'm not sure about International Dawn Chorus Day', Sample says. 'A large gang trooping around together on a dawn chorus walk: it's quite an odd concept. Maybe it spills over from the cultural context of our music: we gather to listen to and participate in a musical performance. I think maybe listening musically to other creatures works better when you're alone. Though I dare say for a newcomer it helps to have someone guide you into the subject.' Only seldom will there be interest from younger listeners, mostly when the subject matter is presented in unusual locations: 'A DJ spot I did in a London shopping centre was very rewarding in this respect. There was a young Asian guy, very smartly dressed in suit and tie. After a while he came over and asked me about the sounds I was playing: he struggled to grasp that these were straight recordings from the natural world. He wanted to know where he could go and hear such amazing choruses and soloists for himself.'

Scientific progress

At the same rate at which public attention is melting away, technological progress is speeding up. Not only have devices become lighter and more affordable, batteries more powerful and microphones more precise. But the technology used for documenting bird song not just in sound, but also written form, has continually improved. Whereas someone like Messiaen would still apply a self-developed shorthand to note down the calls of birds, the advent of the Sona-Graph would allow for their increasingly precise mechanical representation. In his article 'A Brief History of Spectrograms', renowned field recording specialist and editor of *Colorado Birds* magazine Nathan Pieplow, sums up the history of the Sona-Graph from 1951, when the Kay Electric Corporation started producing and selling machines allowing for the translation of audio signals into graphics – or, as the expert would say, audio spectrographic analysis. As Pieplow points out, 'the original Sona-Graph had two settings, "narrow-band" and "wide-band." The narrow-band setting was more accurate in terms of frequency, but less accurate in terms of time, and it tended to create spectrograms that looked rather anaemic, like they had been drawn with a thin pencil. (...) Another common practice was to cover the paper Sonagram with a sheet of acetate and trace the spectrogram with a pen to produce figures for publication.

Many authors preferred this method because it created "cleaner" results – the pen eliminated background noise and equalized levels in the bird's sound. These traced spectrograms were some of the best that could be created in their time, but they were also in some ways insidious: they resulted in a simplification and sometimes a distortion of the sound, as human artistry made "prettier" what the Sona-Graph had rendered. I tend to think that many authors of the day actually traced the spectrograms they published even if they didn't admit to it in their Methods section: for example, the spectrograms of Coutlee look like they may well have been traced.'

In 1967, the first edition of the *Golden Field Guide to Birds of North America* was the first publication on birds to include Sonagrams, an idea which would eventually spill over to Europe, where books like *Die Stimmen der Vögel Europas* (Voices of the birds of Europe) would emulate the example of the *Golden Guide*. Although the book was a mixed success in critical terms and spectrograms remained hard to read for the non-initiated, they would remain the instrument of choice for field recordists. In the mid-90s, according to Pieplow, digital spectrograms took the technology to a new level: 'They can now be instantaneously adjusted to show finer gradations in the relative loudness of parts of the sound, using a grayscale digital display.'

Even more important than these breakthroughs was a change in understanding of how to listen to bird song in the first place. Yannick Dauby, who has appeared in an earlier chapter of this book as an expert on frog sounds, describes this change in understanding with a simple slogan: Human brains are too slow! What Dauby is referring to is the far higher speed at which auditive information is processed by birds compared to humans. Birds are effectively living in a faster world than ours and to hear what they are hearing, we need to slow their singing down. One of the first to grasp and work with this concept was Hungarian biologist and musicologist Peter Szöke. On *The Unknown Music of Birds*, one of the most unique recordings in music history, called 'bizarre' and 'curious' by some and 'revelatory' by others, he first slowed down bird recordings and then asked a soloist from the Budapest opera to sing them. The result is perplexing, offering an entirely new perspective on the acoustic worlds of birds, one filled with a plethora of unsuspected nuances, richness, astounding patterns, rhythms and melodies. As Dauby shows, various composers have, over the years, developed similar approaches, incorporating bird song at different speeds into their oeuvre.

Organs of song

These feats notwithstanding, the act of sound production in birds has remained fiendishly complex. Some birds are capable of singing without interruption for 95 seconds, producing 1,500 different elements over that period. And next to bird song, some species have added instrumental noises into their vocabulary, working with fizzing and hissing, clattering (storks) or clicking (ducks), further adding to the richness of their timbral potential. And yet, the basics of sound production are anything but hard to grasp. The main organ responsible for sound production in birds is the syrinx, situated right at the forking of the windpipe. There are three different syrinx-types, but they all work in a similar way: Since the syrinx is part of the respiratory system, a combination of respiratory muscles and the bird's body – including elements of the skeleton and flexible membranes – is used for creating sounds. The airsack of the collarbone envelops the syrinx and by changing its pressure on the syrinx, birds can directly influence the pitch of the tone. By means of adjusting the pressure of the airstream flowing through the windpipe, they can also influence its volume.

Scientists today have three different theories with regards to how, precisely, these processes work together in creating song. A first model is based on the idea of membrane oscillations against the syrinx – i.e. the bird controlling the air flow and alternating waves and nodes through the windpipe. A second model assumes that flexible parts inside the syrinx can be closed to form a valve, which, in turn, can be opened to allow through controlled amounts of air. Finally, whistles can be created by the narrowing of certain passages in the syrinx to the size of slits. Making things just slightly more complicated, the latter factors are in turn dependent on minute anatomical differences between different species or even different techniques. What's more, some species may not exclusively be using one of these techniques, but all three of them. And then, there is a natural limit to just how much we can understand: Since the larynx is covered from sight almost completely by the windsack, it has been impossible so far to actually take a look inside to see what is really happening in the act of singing. This has made it exceedingly hard to arrive at satisfying conclusions.

In their seminal work *Ornithologie*, Einhard Bezzel and Roland Prinzinger maintain that there are seven fundamental types of calls, with some species working with up to 350 different motivic categories. How do birds learn them? For one, singing is a combination of a genetically inherited basic amount of song patterns and a phase of learning as part of a group – Kaspar-Hauser-birds raised in perfect isolation will never be able to develop their species' full repertory. The first, most important phase, begins almost immediately after birds have fledged. Within the first year of their life, young chaffinches are capable of collecting so much information that they can go from their first, feeble croaks to almost fully formed bird song. Other species take longer to attain the full vocabulary, while yet others, including canaries, are believed to continue adding new calls to their song throughout their lives. To stop the young birds from just adding everything they hear, there must be some filters allowing through only specific information – in some cases, these filters may be shaped by the basic song patterns already present at birth, in others, they may be provided by a close learning relationship with their father. Nonetheless, there are numerous cases of mixed singing, where one species includes or imitates elements of a different one's chants into their own. And then, of course, there are dialects, regional variations of a particular song structure, some of which are so minute that they can only be demonstrated by the use of a sonagram. There are different hypotheses regarding the reasons for their development and effects. One of the most fascinating of these assumes that regional variations can lead to a restriction of gene flow to only local individuals and thus serve as a first step in the development of new species with unique songs and sounds.

A question of meaning

What are all these different calls and songs for? When it comes to calls, Bezzel and Prinzinger list goals like demonstrating aggression, alerting others of a predator, peaceful behaviour between male and female birds in a partnership or within a group, behaviour of young birds among themselves and towards their parents (and vice versa) as well as a variety of special forms such as echolocation.

When it comes to bird song, Geoff Sample explains that 'song conveys species identity, sometimes clan affiliation, individual character and "quality" and in many cases, whether a male has a mate or not, the level of lust and readiness to challenge another male. But often the communication resides not in the song itself but in the performance, the mode of singing and the interaction – for instance whether an individual continues to sing when another singing male approaches; singing provides a way for birds to negotiate space and status, to solicit sex, to impress other individuals, without resorting to physical violence. And then there are so many instances where birds sing that don't fall into the simple model. Subsong, where a bird seems to be practising, exercising its repertoire or developing its voice. Flock song where it seems to function as a bonding of the individuals – like a congregation (Latin: con – together, grex/gregis – flock) singing hymns together.' He stresses that it can be misleading to think of bird song as a homogeneous activity and that it is decidedly not just a male practise, nor confined to the breeding season. And most of all, he emphasises that there can never be a unified theory, simply because birds are so different: 'Corn bunting males sing stereotyped or formal songs and, in undisturbed habitats, tend to form dialect groups; the form of the song becomes like a gang symbol. Sedge warblers on the other hand have elaborate improvisatory songs and it seems that females choose males largely on the basis of their song. What's more, Dr. Clive Catchpole found that males with more versatile songs, built from a wider repertoire of elements, gained mates before those with simpler songs. Sedge warblers have a monogamous breeding system where the males actively contribute to rearing their offspring, so here obviously the quality of the male is important for a female and it seems that song is an honest indicator of male quality. The complexity of male song in sedge warblers is being driven (in evolutionary terms) by female choice.'

This also explains why Sample thinks there is still a long way to go. Science may have established the biological and behavioural context of bird songs, a 'framework for understanding what it's about'. And yet, how to join up the dots is still open to interpretation. For him, the most revealing insights are most likely to be found somewhere at the cusp between traditional bioacoustics and musicology. He dotes on the recent discovery that certain complex sounds, which can only be learned by many years of trial-and-error and experience, are particularly attractive to female canaries. Dubbed 'sexy syllables', they could point towards a deeper insight into what makes a particular song successful. But as far as the insights may go, the search can never be over: 'I often ask myself, if, say, marsh warblers are extinct in 100 years' time, or the nature of our moorlands has changed, have I made a recording that does full justice to what they can sound like, the power of their voices and richness of their songs, or the scenes I've heard in my lifetime? The answer is rarely yes, and that keeps me going. I recorded numerous marsh warblers over a period of years before I found a singer that really bore out a description

I once read - "intensely passionate, even to the verge of delirium" (John Walpole-Bond). Those people who say, "well surely you've recorded every British species now?" are missing the point.'

Kevin J Zimmer, 'Ted Parker Remembered' in Birding XXV (6): pp. 377–380, 1993

Tomas Carlberg, 'Ted Parker: the Man, the Myth and the Legend' in Fauna & Flora 1/2011

Don Stap, 'A Parrot Without a Name: The Search for the Last Unknown Birds on Earth' in Knopf: pp. 104–160, 1990

Nathan Pieplow, 'A Brief History of Spectrograms' http://earbirding.com/blog/archives/1229

Einhard Bezzel and Roland Prinzinger: Ornithologie, Eugen Ulmer, 1990

THE CAT'S MEOW: CATS PART I

Cats may be holding on to their juvenile meow to manipulate us.

It's something that cat owners have known instinctively since they entered into that special understanding with their feline friends; that uncanny ability of the house cat to communicate with its owners. And now, science can prove that our instincts are correct. Scientists believe that cats have developed an anthropogenic means of communication that appeals to our auditory and emotional responses.

Researchers have long been fascinated with feline vocalisation. In 1944 Dr Mildred Moelk wrote a seminal paper in *The American Journal of Psychology* that documents and describes in great detail the sounds that domestic cats produce. She categorised the sounds into three groups; cat-goal, cat-man and cat-cat. The cat-goal group describes the noises that a cat makes when it wants something. The cat-man group vocalisations arise when a cat's attention is focused on a human in a friendly, petting or sympathetic manner. The cat-cat category is regulated mostly by the life-cycle of the cat including the meow of the kitten and the caterwaul of mating season.

Moelk also groups the basic vocalisations into three categories, which relate to the position of the mouth. There are the murmurs of the closed-mouth variety, the open and closing mouth of the meow and finally the sounds produced when the mouth is held open in a fixed position such as growls and chatters.

Moelk effectively demonstrates that cats do indeed have different patterns of communication for different scenarios and you could say that, in their own way, cats 'talk' to us. Cats are born with the ability to vocally express themselves in a variety of ways which appear to be related to their level of satisfaction or dissatisfaction. The house cats' dependence on their owners has led them to expand upon their vocal repertoire in order to better let their human friends know how they're feeling. This domestic vocal repertoire is not shared by their wild relatives.

French art critic, author of *Les chats* and cat-lover Champfluery (a.k.a Jules François Felix Fleury-Husson), counted 63 notes in a cat's vocal stylings. They can express complicated vowel sounds, as Moelk aptly details in her essay, and up to nine consonants. Author and academic Dr Nicholas Nicastro documented 535 cat meows over a span of 36 hours among 12 domestic cats, in his popular dissertation that suggested cats have learned how to manipulate their owners.

It is well established that while only the *Felis* genus – or small cats – can purr and only the *Panthera* genus – or big cats – can roar, all felines can meow. Even today, the exact mechanics of cat vocalisations elude scientists. Recent studies report unknowable physical elements of vocal tracts and counter-intuitive findings related to size and frequency. No one is really sure how a cat produces its purr, though it is thought to relate to the hyoid bone in the throat. In the first comparative and quantitative report of purring in domestic cats, veterinarian anatomist Dr Gerald Weissengruber suggests it has something to do with vocal folds, that the sound is made by rapid twitches of the vocalis muscle. In big cats there are large pads within the vocal folds which would make it hard to contract with any rapidity and therefore make purring difficult.

But it seems that while we can't pin down the 'hows' of feline vocalisation, we can make assumptions about the 'whys'. In his 2004 essay, 'Perceptual and Acoustic Evidence for Species-Level Differences in Meow Vocalizations by Domestic Cats (*Felis catus*) and African Wild Cats (*Felis silvestris lybica*)', Dr Nicholas Nicastro studied vocalisations of the domestic cat in comparison to those of its closest wild relative, the African wild cat. They recorded 535 various cat meows, from both domestic and wild cats. They played these recordings to a number of human test subjects and assessed the responses of listeners with regard to acoustical characteristics and perception. In every instance, the candidates considered the domestic cat vocalisations far more pleasant than those of the wild cats.

Nicastro aimed to prove that domestic cats have developed vocalisations that appeal to humans and solicit sympathy and response. The meow is a sound made by kittens, across all felids both wild and domestic. However, it is generally only the house cat that continues to meow into adulthood. It's been suggested that this juvenile behaviour indicates the arrested development of domestic cats. It also indicates that cats figure out that we much prefer the sweeter, higher sounds of the meow to the more guttural utterances of an adult cat. It's been noted among animal rescue workers that the meow is largely absent in feral adult cats, and yet after regular contact with human caretakers, feral cats will develop a meow.

Nicastro chose the African wild cat for the study because it is one of the rare example of wild cats that do meow in adulthood. He could therefore play recordings of both wild and domestic meows to the listeners with the assurance that the African wild cat's meow was in no way designed to appeal to human ears. After his analysis of the listener responses, Nicastro reported that domestic cat vocalisations are indeed more effective at exploiting human response. Compared to the deeper meows of the wild cats, the domestic cat meows are higher pitched which taps into the human tendency to respond to higher, less threatening and ultimately more appealing tones.

Research carried out by Dr Karen McComb in 2009 at Sussex University examined how some domestic cats embed certain tones and intensities in their purrs to garner a more immediate human response. Most recently in 2012, phonetics expert Dr Susanne Schötz at Lund University carried out a pilot study to speculate that cats mimic paralinguistic – perhaps even linguistic – information by varying their fundamental frequency. In her study Dr. Schötz suggested that the extraordinary variation in cat vocalisations might point to the existence of, more or less, a cat language. She thinks their sophisticated ability to modulate the frequencies and combination of their meows and murmurs might have a purpose. That purpose, she hypothesises, is a fine and complex method of vocal communication. Her findings indicate, as did Dr Nicastro's, that cats have different vocal communications depending on whether they are communicating cat-cat or cat-human. These findings hint at a level of linguistic sophistication that has long been appreciated by cat lovers and only recently accepted by scientists.

Since humankind's intimacy with cats began, we have noticed the extraordinary expressions of these mysterious creatures and wondered about their ability to manipulate our emotions. Whether they're protesting about the brand of cat food on offer or being indignant about a visit to the vet, we have always understood what they are trying to 'tell' us. Science, as always, takes a while to catch up with instinct. And it seems this time we were right about our feline friends. Recent research has vindicated our suspicions with examples of how cats moderate and change their natural, wilder vocal tendencies to produce softer, higher meows and murmurs that please our ears and tug at our heart strings.

Susanne Schötz & Robert Eklund, 'A comparative acoustic analysis of purring in four cats' in Quarterly Progress and Status Report TMH-QPSR, Volume 51, 2011. Proceedings from Fonetik 2011, Royal Institute of Technology, Stockholm, Sweden, 8–10 June 2011, pp. 9–12

Robert Eklund & Gustav Peters, 'A comparative acoustic analysis of purring in juvenile, subadult and adult cheetahs' in Proceedings of Fonetik 2013, the XXVIth Swedish Phonetics Conference, Studies in Language and Culture, no. 21, 12–13 June 2013, Linköping University, Linköping, Sweden

Susanne Schötz, 'A phonetic pilot study of vocalisations in three cats' in Proceedings of Fonetik 2012, Department of Philosophy, Linguistics and Theory of Science, University of Gothenburg

Dr Nicholas Nicastro, 'Perceptual and Acoustic Evidence for Species-Level Differences in Meow Vocalizations by Domestic Cats (*Felis catus*) and African Wild Cats (*Felis silvestris lybica*)' essay, Cornell University, 2004

Carl Van Vechten, 'The Cat in Music' in The Musical Quarterly, Vol. 6, No. 4 (Oct, 1920), pp. 573-585, Oxford University Press

THE POWER OF THE PURR: CATS PART II

The cat's purr is a sound that can literally save lives.

It's an old saying in the veterinarian trade that if you put a cat in the same room with a bunch of broken bones, the bones will heal; and we've all heard the saying that a cat has nine lives. Aphorisms like this are often relegated to urban myth and legend, but it looks as if science has found some truth in these old sayings. Anyone who has ever had a pet cat will be familiar with that unique feline quirk, the purr. Until quite recently it was thought that anything from rabbits, pigs and even elephants could purr. But studies have shown that it is only the cat, or Felidae species and possibly two kinds of genets from the Viverridae species, that actually purr – in the strictest sense.

A purr is defined by Dr Gustav Peters as a 'continuous sound production [that] must alternate between pulmonic egressive and ingressive airstream' (2002). He qualified it further by suggesting that it can go on for many minutes at a time. Essentially, it means that only cats and some genets can make the sound we all recognise as purring while they breath in and out, resulting in a continuous, rhythmic vibration.

The physics of the purr are a bit of a mystery to scientists, because cats don't possess a unique anatomical feature that is responsible for the sound. But it seems that purring is facilitated by a fully ossified hyoid which exists in most cats of the *Felis* genus. The big cats of the *Panthera* genus include the tiger, lion, jaguar and leopard, and they cannot purr in the strictest sense. They can produce a similar sound, but only when exhaling. It is thought that this is because their hyoid is only partially ossified, a physical attribute that also enables them to roar. This distinction between roaring and non-roaring cats was suggested all the way back in 1834 by biologist and comparative anatomist Richard Owen, who would have used the differences in the hyoid to distinguish between the roaring and non-roaring cats. But to confuse things even further, there is an exception to this general rule. The snow leopard possesses a hyoid that is also not completely ossified and yet it can purr.

So even though scientists are not really sure how it's done, they seem to have come to an agreement about what defines the purr. A domestic cat purrs at a frequency of 25 to 150 vibrations per second. With differences between cats and also between the ingressive and egressive output, we could generally say that they purr between 20-27 Hz.

It was once thought that a purr was indicative of a contented cat. And while this is certainly true, it is also true that cats purr when they are nervous, in pain, ill or asleep or even just to solicit a response – from both other cats and humans. Researchers now believe that the purr is linked with survival. It has been recorded in veterinarian science journals that cats have an uncanny knack of surviving falls from very high places, the highest reportedly being 45 storeys. Cats recover more quickly from bone injuries than dogs, suffer fewer complications from surgery and have less orthopaedic and muscular trauma. It is now thought that the cat's ability to heal itself so efficiently is due to its purr.

Medical science has established that bone strength can be improved by up to 20 per cent when treated with frequencies of around 20-50 Hz. Higher frequencies, closer to 120 Hz, can promote healing in tendons and ligaments. The cat's purr is also within the range of frequencies known to relieve chronic and acute pain. Vibrational stimulation is known to relieve pain, promote healing, bone and tissue strength and improve circulation and oxygenation. This relatively new discovery could mean that we can reverse the effects of conditions that deteriorate bones like osteoporosis and provide complementary health care for injuries and pain management.

So it seems that our urban mythology about cats and their seemingly magical ability to survive was spot on. Maybe the ancient Egyptians weren't so unfounded in their worship of this creature. Empirical evidence has shown that biologically speaking, the purr has survived for a reason. The cat's in-built healing system can not only benefit themselves, it can also benefit others.

Eklund, Robert, Gustav Peters & Elizabeth D. Duthie. 2010. An acoustic analysis of purring in the cheetah (Acinonyx jubatus) and in the domestic cat (Felis catus).In: Proceedings of Fonetik 2010, Lund University, 2–4 June 2010, Lund, Sweden, pp. 17–22

Leslie A Lyons, Scientific American, April 2006, http://www.scientificamerican.com/article/why-docats-purr/

Elizabeth von Muggenthaler, 'The Felid Purr; A healing mechanism?', in the proceedings from the 12th International Conference on Low Frequency Noise and Vibration and its Control held in Bristol, UK, 18th to 20th September 2006. http://www.animalvoice.com/catpur.htm

CLICK. CLICK. TICK: BATS

Echolocation is a stunningly inventive navigation tool.
But is it also a language?

Lena Grosche has a thing for bats. Ever since her days as a student of landscape ecology, she has made them her focus of attention and today co-runs Echolot, Germany's leading agency for bat sciences. Whenever big infrastructural projects or construction work are being planned, decision takers will call Grosche to explore the impact on the bat population. And whenever that happens, she'll leave the house in the middle of the night with a team of colleagues and volunteers to catch a few specimens, establish data about their development and to record their chants – one of the most unique and fascinating facets of animal song on the planet.

Bats use echolocation for navigation, but they're not the only animals to do so. Dolphins and whales in particular, as well as some rare bird species, also use more or less sophisticated forms of it. But owing to their elusive character, somewhat frightening appearance and a flowery mythology, bats have always been able to inspire the human imagination just a little bit more. For centuries, it remained unknown how they were able to navigate at night. After all, although bats aren't actually blind, they do have rather limited eyesight. As far back as 1790, Italian priest, biologist and physiologist Lazzaro Spallanzani had come within arm's length of solving the riddle. Spallanzani locked up an owl and a bat in a perfectly black room to observe their capacity of flying in complete darkness. While the owl would constantly bump into different objects blocking its flight path, the bat managed to stay clear of them. After a letter describing his inquiries had fallen into the hands of Swiss scientist Charles Jurine, another experiment was added to the equation: After plugging one of the bat's ears, it was no longer able to avoid the occasional collision. His conclusion that bats were equipped with a 'sixth sense' and required their sense of hearing for spatialisation would prove to be correct – but Spallanzani did not yet have the means required to translate his suppositions into hard facts. It would take a full

140 years and until the arrival of a seminal figure in animal sciences, Donald Redfield Griffin, to make that happen.

More than a reflex

It was clear right from the start that Griffin was different from his colleagues. Born in New York in 1915, he was already a professor by 1942, going on to build a career that would take him to Cornell, Harvard and, finally, Rockefeller. What mainly distinguished Griffin from an earlier generation of zoologists was firstly, that he, as Raghuram/ Marimuthu put it in their essay 'The Discovery of Echolocation', 'challenged the dogma that animals are mindless automatons, controlled solely by instinct and reflex.' Secondly, he would make use of all the innovative technology he could lay his hands on to prove it, going on to make groundbreaking discoveries in the field of bird migration and the hearing facilities of fish. But nowhere would his impact be felt more strongly than in the field of echolocation.

Griffin was intrigued by two articles which had appeared within close temporal proximity of each other after the Titanic disaster. The sinking of the ship had started a race for ideas to prevent a catastrophe like it from ever happening again and in 1912, Sir Hiram Maxim suggested that one might perhaps be able to make use of the bat's as-yet unknown 'sixth sense'. His hypothesis consisted in the idea that 'bats detect obstacles by feeling reflections of the low frequency sounds caused by their wing beats (appx. 15 Hz) and ships could be protected by collisions with icebergs or other ships by the installation of an apparatus to generate powerful sound and a detection device to receive the returning echoes'. Five years later, Hamilton Hartridge proposed the exact opposite, namely that bats might be using sounds of high frequency and short wavelength. Griffin was more convinced by the latter theory and would use it as a departure point for his own inquiries. It helped that physicist George Washington Pierce had already built a device capable of detecting these ultrasonic sounds, although he had actually constructed it for recording cicadas. It would take a string of experiments conducted between 1938 and 1940 until Griffin was able to fully grasp the mechanism at work –

and he would pursue refining and detailing his discoveries for many years after that.

In an essay for *Scientific American*, Professor Alain Van Ryckegham has summed up Griffin's findings: 'Most bats produce echolocation sounds by contracting their larynx (voice box). A few species, though, click their tongues. These sounds are generally emitted through the mouth, but Horseshoe bats (*Rhinolophidae*) and Old World leaf-nosed bats (*Hipposideridae*) emit their echolocation calls through their nostrils: there they have basal fleshy horseshoe or leaf-like structures that are well-adapted to function as megaphones. Echolocation calls are usually ultrasonic – ranging in frequency from 20 to 200 kiloHz (kHz), whereas human hearing normally tops out at around 20 kHz. Even so, we can hear echolocation clicks from some bats, such as the Spotted bat (*Euderma maculatum*). These noises resemble the sounds made by hitting two round pebbles together', Van Ryckegham writes. 'In terms of pitch, bats produce echolocation calls with both constant frequencies (CF calls) and varying frequencies that are frequently modulated (FM calls). Most bats produce a complicated sequence of calls, combining CF and FM components. Although low frequency sound travels further than high-frequency sound, calls at higher frequencies give the bats more detailed information – such as size, range, position, speed and direction of a prey's flight. Thus, these sounds are used more often.'

Not only are bats able to produce astoundingly loud sounds of up to 120 db, which is 'louder than a smoke detector 10 centimeters from your ear', as Van Ryckegham notes with palpable admiration, they are also able to make astounding distinctions when processing the information. According to the author, 'some Horseshoe bats can detect differences as slight as .0001 kHz'. It is through this combination of remarkable talents that bats are capable of navigating so precisely in the absence of daylight.

Bat detectors

What can also be concluded from these elucidations is that it is technically possible to perceive bat calls and clicks with bare ears – if standing close enough to the bats and, with regards to the directionality of the ultrasounds, in their flight path. In fact, without being aware of

Griffin's experiments, Dutch scientist Sven Dijkgraaf had arrived at the very same conclusions just by listening to bats, coining the term 'ticklaut' for their clicks. In practise, meanwhile, most researchers today are not using their ears, but bat detectors, modern versions of Pierce's ultrasound apparatus, to gather the necessary data. Lena Grosche, for example, is using both portable detectors as well as more advanced stationary devices, which generate high-resolution recordings and even allow for visualisations of the echolocation information – although to Grosche, echolocation is purely an acoustic phenomenon and any graphic representation is merely an attempt to translate it to the more visually oriented human world. Japanese field recordist Eisuke Yanagisawa prefers a detector of the heterodyne-type – which is relatively cheap and simply pitches down the bat calls to a frequency range accessible to human hearing in real-time – because 'it has good sensitivity, is easy to use and the least expensive.

It seems each model has a different sound character even in the same types of detectors. I understand that the heterodyne bat detector can't keep the original waveform as time expansion ones do, but I don't think the one I use completely alters the nature of the original source sound. If you listen to cicada recordings, for example, you can hear the common characteristic sounds as we hear in the audible range. And to a greater or lesser extent, recorded sounds are "interpretation" of real sounds even if you record with the "most accurate" microphones.'

So what is communicated through these sounds? According to Grosche, quite a lot, since bats have an astounding expressiveness. There are many different social calls used for interaction, although their specific content remains to be decoded. Calls used for mating, between mother and child, as well as for securing one's territory have also been identified – in human terms, they're communicating, as she puts it, information like 'Get lost, this is my territory' or 'Come over here, you'd be perfect for my offspring'. A recent study has also suggested that tropical bats may be sending out welcome calls to each other and to confirm that they're of the same species. But how much creativity does she see in these calls? 'I wouldn't say bats are creative per se. Neither have I heard of individually developed call techniques.

In fact, I'm sure the techniques currently in use were developed through long evolutionary phases. On the other hand, different bat species have developed differentiated and highly specialised location systems. And just as with probably every mammal, each bat has a voice of their own. There is, for example, a certain variability within the main frequencies of a particular species. There is also some research being done into geographically different dialects.'

The latter facts may sound hopeful, yet one should never forget that echolocation significantly depends on environmental conditions. As long as there is no established method of filtering these out from the equation, extracting the communicative aspects will remain exceedingly hard. If two bats are emitting different calls, for example, they may be trying to communicate almost the same message, but adapting their calls to different situations. Even humidity can, as Grosche points out, play a role in this. Seventy years after William Griffin's seminal discoveries, the mysteries of bat song are still far from completely resolved.

---

http://www.scientificamerican.com/article.cfm?id=how-do-bats-echolocate-an

http://www.ecoobs.com

http://www.batcon.org

Donald Redfield Griffin, 'The Discovery of Echolocation' in Resonance Journal, Feburary 2005

Donald Redfield Griffin and Robert Galambos, 'The Sensory Basis of Obstacle Avoidance by Flying Bats' in Journal of Experimental Zoology Volume 86, Issue 3, pp. 481–506, April 1941

LEARNING TO LISTEN

How do animals hear & how should we listen to them?

SPECIALISED TOUCH: HOW DO ANIMALS HEAR?

Ears are great tools for hearing – but they're not the only ones.

Cat 'owners' will gladly testify to the popular saying that 'a cat's hearing apparatus is built to allow the human voice to easily go in one ear and out the other.' To scientists dealing with animal sounds and vocalisations, meanwhile, the quip actually contains quite an essential insight: That a meaningful analysis of animal song can never be complete without an analysis of their hearing. This is so obvious that it should seem surprising at first to find that progress in this field has so far been rather unimpressive. And yet, there is a very good reason for the striking lack of reliable facts: Measuring animal hearing is exceedingly hard. One method consists in conditioning animals to respond to auditory cues in a conditioned way, for example, by selecting the right dispenser offering a reward. Another, the so-called brainstem auditory evoked response (BAER), measures brainwave activity by means of three electrodes and a simple headphone, through which sound is played to a non-human proband. Both, however, offer limited information at best.

And so, most accounts of the topic typically resort to the kind of 'celebration of extremes'. Typically, these tend to praise the supposedly far superior hearing capacities of animals: The greater wax moth is capable of sensing sound frequencies of up to 300 kHz, for example, 'the highest recorded frequency sensitivity of any animal in the natural world'. The roar of lions, too, is said to contain infrasound portions, which are used to stun their prey. Elephants, on the other hand, can detect subsonics, which can travel across huge distances – which may also allow them, as some have suspected, to detect approaching thunderstorms or quakes long before they strike. Yet another source informs us that 'insects have tympanal organs that work as well as ears, and in fact give them far better hearing than humans.' In reality, when taking audible bandwidth as a criterion, human hearing doesn't compare quite as badly to those of other species as is often claimed. We can't hear the ultra-low frequencies accessible to an elephant, but we can hear considerably higher ones. And while our ears are incapable of processing high-pitch information the way a bat could, their conscious hearing is already cut off at 2,000 Hz on the lower end of the spectrum, where humans can still process a wealth of information. In the big picture, humans neither have the most refined pair of ears – nor do they have the worst.

These relative advantages and disadvantages are quite obviously the evolutionary result of adapting to very specific demands and conditions. The reasons why cats have developed a remarkable sensitivity to higher frequencies without losing their lower-end hearing is that their main prey, mice, communicate precisely at these frequencies. Also, thanks to her excellent hearing, a mother cat can locate and retrieve a lost kitten. Their hearing, in fact, is so precise that a blind cat will often be perfectly capable of moving around the house on the strength of her ears alone without constantly bumping into objects. What's important to note is that this sense of hearing isn't just down to superior signal processing in the brain, but begins right at the start of the signal chain: 'A cat can separately rotate the outer ear flaps (pinna) 180 degrees, each functioning like a mini-satellite dish that retrieves data to be analyzed by the cat's brain. The pinna collect sound and amplify it,' says Dr Weigner (a board certified specialist in Feline Practice and Diplomate of The American Board of Veterinary Practitioners who founded The Cat Doctor in Atlanta, Georgia). The delay in time that it takes for sound to reach one ear versus the other helps them pinpoint the source. 'They can differentiate sound direction from a meter away to within 8 centimeters,' says Dr Weigner, 'which makes a difference in locating prey.'

Alternative approaches

While comparisons between cats and humans are still fairly straight-forward, things quickly become rather complex as soon as one turns towards other species. The more one immerses oneself in the topic, the more it becomes apparent that the broad term 'hearing' covers a diversity of different approaches and that our understanding of the phenomenon of sound and of hearing as such is still rather limited. As percussionist Evelyn Glennie has strikingly pointed out: 'Hearing is basically a specialised form of touch. Sound is simply vibrating air which the ear picks up and converts to electrical signals, which are then interpreted by the brain. The sense of hearing is not the only sense that can do this, touch can do this too. If you are standing by the road and a large truck goes by, do you hear or feel the vibration? The answer is both. With very low frequency vibration the ear starts becoming inefficient and the rest of the body's sense of touch starts to take over. For some reason we tend to make a distinction between hearing a sound and feeling a vibration, in reality they are the same thing.' No wonder the animal kingdom has come up with so many different takes on how to process them. Elephants use their feet to pick up low frequencies. Spiders and cockroaches have tiny hairs on their legs for sound wave detection, caterpillars across the surface of their body. Birds, on the other hand, appear to hear not just with their ears, but, according to Philip Radford, by using the so-called 'corpuscles of Herbst. These touch-type nerve endings (...) are to be found in the follicles of feathers, in the deep tissues of the leg and in the beak. Birds are said to respond to distant low-frequency sound, such as gunfire, which is inaudible to humans. It may well be that the corpuscles of Herbst, rather than the ears, are the receptor organs in such instances.'

To understand what animals are saying, therefore, we must understand what and how they are hearing. Since cats are so particularly susceptible to higher frequencies, perhaps the human voice might not travel quite so easily in one ear and out the other if you were speaking to yours with a higher voice. Even though, as Dr Weigner admits, 'the response may depend on how sleepy he is.'

Aubrey A Webb, Brainstem auditory evoked response (BAER) testing in animals http://www.ncbi.nlm.nih.gov/pmc/articles/PMC2643461/

http://phys.org/news/2013-05-world-extreme-animal.html

Betty Vosters-Kemp, How Does Hearing Work in Fish, Caterpillars and Other Animals? http://www.avalonhearing.com/tag/hearing-in-animals/

Karen Commings, Cats' Sense of Hearing, Catwatch 2007

Evelyn Glennie, Hearing Essay, 1993:
http://issuu.com/evelynglennie/docs/hearing_essay__revised_2015_

LISTENING TO ANIMALS: AN INTERVIEW WITH CHRIS WATSON

*One of the world's leading field recordists, Chris Watson is
constantly thinking about how best to listen to animals. Still, he
opines, our understanding of it can be limited at best.*

**Over the years, you've recorded a wide range of animal sounds
from all over the world, from very 'small' sounds to very 'big' and
loud ones, from familiar noises to entirely alien ones. Can you try
and describe why listening to animal sounds is so absorbing to you?**
First of all I don't use the word 'noise' in that context, I'll use the word
'sound'. I think it's very important that we make a distinction between
the word 'noise' and 'sound'. The vocabulary for sound is unfortunately
very limited, we use a lot of visually oriented phrases so I'm very careful
about that distinction. When I was in a band, I used to work a lot in
the studio making music and, like I do now, I used to take walks and
go outside, because it's a good way to give your ears a rest and consider
and listen. When I started doing that carefully, and this time it was out
in Derbyshire, Sheffield where I grew up, I suddenly realised I became
more interested in the sounds I was hearing outside than those we were
trying to create inside.

As a teenager I was interested in musique concrete and I became
fascinated by some of the contemporary composers of that time like
Karlheinz Stockhausen, Olivier Messiaen and Pierre Schaeffer and I
really got immersed in that. I realised that what I was listening to could
be regarded in a musical way and could be used as tools for composition
in a more satisfactory way than I could contrive in the studio. Then I
became more interested in the idea of bioacoustics and recording animal
sounds. The natural world, birds of course in particular, are the most
vocal so it's a good place to start. So I began recording wildlife sounds.
I've always been interested in their habitats and environments as well. I
believe they have a particularly interesting voice and speak for themselves
in interesting ways that I'm very keen to convey through my work.

**Before pressing record, you first have to select a subject. How does
this selection phase work for you?**

The older I get, the less recording I do and the more listening I do. I'm becoming more and more selective and I think that's a process simply down to experience. I don't think you can short circuit it really. So I go to work and I listen and I find something that interests or engages me. If I can't find anything like that it's just not worth it.

I've recorded so many hours of rubbish and so I'm very careful now about what I record. I often research before I go somewhere, whether it's the moorlands, the arctic or downtown Newcastle. I'll go at the right season or at the right time when I think there might be something interesting to record. I quite often work to a kind of score or map of the sounds that I want, or think I want to obtain. Although of course one of the exciting things about working on location is that you have to be prepared for those wonderful chance moments. That's part of it for me, I like to plan but I also like to keep myself open to the potential for unique events which often happens. So I go and I listen and then quite often I make repeated visits. Usually, in some of the more remote locations, I use very long cables and I plant microphones and I mike up areas and then run cables back to a place where I can listen without affecting any behaviour or disturbing anything. Then I record selectively. When I'm doing my workshops I try to convey exactly the same thing. You can't tell people not to record. What I'm trying to do is get them to engage with this idea of listening creatively and then use their acquired skills and judgement to make a decision about when to record. Otherwise you just end up with hours and hours of stuff that's not very good and that can be a very disheartening part of the introduction to the post production process.

What kind of sounds will interest you?
Anything that's good, really. Because of the demands on my time I rarely just go off and record something. Nearly everything I do is planned, even if it's a weekend trip into Northumberland to get a sound I want to record for something else, I plan what I do.

With the Tyne River *Going with the Flow* project, I became really interested in the mechanical sounds lower down on the river. For instance, I recorded a scrap yard beside this amazing car crushing device. I recorded the Tyne rowing club with these people rowing up and down on the river and got these great rhythms and sense of the oars entering the water. I recorded the mechanical sounds of the ferry and the people and the ship's horns and I got really immersed in that environment. I can usually find things in any location, but it has to engage me. It doesn't always have to be the natural world.

Last year, Touch released my album *El Tren Fantasma*, which recorded the sounds of a railway journey across Mexico. Since then, and including that project, I've become really interested in railway sounds, the constant rhythm and harmonics you can hear on the railway.

I lived and slept on that train, we were on it for five weeks. I don't know if you've ever slept on a train but it's very womblike. You're in this environment which is vibrating and has a rhythm and I'm sure it's got something to do with the appeal. The pulse of railway is similar to a heart beat and that must be one of the first sounds we hear before we're even born. We're 16 weeks in our mother's womb when we start to hear or are exposed to the world of sound, diffused through amniotic fluid. It's my theory that we find sounds like the railway sounds so absorbing because it replicates a heartbeat.

When in the studio, how do you edit your recordings down to a suitable length for an album track – say, if you've just recorded a long passage involving animal song?
This idea of listening is really interesting and it's one that I spend a lot of time thinking about. If I'm making a CD then I'm limited by the length of the media, which is about an hour. But if I'm doing installations I have a different time scale. With film soundtracks, sometimes I've only got seconds. And so a lot of the time I am constrained by the media with which I'm working.

In terms of installations where I've got more of a natural span. 'Whispering in the Leaves' came out at 16 minutes and I didn't contrive to do that, I just felt as if that time frame suited the material. Sunrise in a tropical rainforest is a very rapid occurrence, because close to the equator the sun rises and sets quickly and there is a very short dawn chorus. Whereas the dawn chorus up here in our Northern latitudes can last for up to 90 minutes. And so I made something that's got the integrity of the piece from where it was recorded but also something that's listenable. You have to engage your audience or you're wasting your time.

Recently I was talking to a group of people called The Quiet Club and I've had conversations with the people from The Wired Lab in Australia about the concept of extended listening. There are people that are prepared to listen to something for extended periods, sometimes several hours. It's quite exciting that there's this growing culture of people who are happy to engage with sound or music, call it what you will, for something outside of the typical iTunes song, something that lasts not for minutes but for hours.

The Quiet Club do these performances which last up to two and a half hours. I've just released a fundraising piece on Framework radio that is a two and a half hour recording I made in the rainforest in Borneo last year. That's an extended listening piece and so I'll be interested to hear what happens to that.

Has the possibility of manipulating recordings to better reflect how animals listen to their own calls changed the way you listen to animals yourself?
I take issue with the statement that there's a possibility to collect recordings to better reflect how animals hear. We have this very arrogant notion that we're top dog in the animal world. We're patently not. We can't understand the sense of communication in the animal world, we can dip into it, in our very restricted frequency range, but we really have no idea what's going on. I think it's arrogant to suggest that by simply manipulating recordings we can understand them. We can reveal things in a different way, but I don't think we can reveal anything that the animals don't already know.

But do you feel as though your work as an artist is part of an effort to arrive at a better understanding of other species?
I'm interested in getting people to listen to animals and their environment. Not only to understand but to listen to the world around them. Often it's much more interesting than you would first imagine.

Whether it's a back garden in suburban Newcastle or the middle of the desert. A fundamental part of my work is to try and recreate that sense of excitement of being out there, listening to things in the real world and being able to come back and present it in a way that people can relate to. That's the great thing about sound; it doesn't need any great artistic justification. People get sound very directly. It strikes into our hearts and imaginations in a very unique way.

When you once spoke about the purr of a cat, you mentioned that you love zooming in on the sounds of a particular animal, because they reveal something about them. From your experience and speaking more concretely, what is it that they're revealing?
I don't think the cats are revealing anything about themselves. What I was probably saying, not very eloquently, was that the recordings reveal to me the beauty and the depth of some of those sounds and maybe we're hearing them rather more like the way the animals appreciate it. I find it very stimulating creatively to listen to those sounds because we're hearing it in a new way. You can hear patterns and rhythm and depth in a way that you can't perceive literally a few metres away. So it reveals to me a better appreciation and understanding of the animal I think. You get a greater notion of the incredible complexity of these sounds.

How has the way you listen to animals changed over the years, owing to your increased experience in the field?
Technology, hand in hand with listening, is probably what's changed most with me over the last 25 years. I've become a lot more selective because I listen more and record less. That helps me both to further my understanding of what I'm listening to and how I want to work with it. The other thing that's gone hand in hand with that over the last 25 years is improvements in technology. I don't have to listen to my recordings through the blur and patina of noise from analogue recordings, which really masked a lot of quiet, transient sounds, tape hiss in particular. Now with digital recording and high sample rates, I can hear something in the studio and it sounds exactly the way I remember hearing it in the field. That is spectacularly good for me. It's amazing to bring those sounds

back to the studio and to work with them. Again, a step up from that will be to work spatially with that sound, which we're getting closer to.

What's your ideal state of mind for listening?
Listening and recording is a solitary experience. It's impossible to do it in a group. I have to be alone, and focused and comfortable, so I can listen for extended periods, take notes and record.

Like my distinction between sound and noise, I make a great distinction between hearing and listening. We hear everything and spend much of our day filtering out all the crap in order to get to the message. But when we can go somewhere and open our ears and listen and actually engage with a place, for me, that's a very creative and stimulating process. It opens my imagination up to the potential of places and animals and sounds.

On location it's a very solitary process, but conversely, when you're presenting work to an audience there's a bit of an obstacle there. I'm still working on presenting my work in a way that's satisfactory to me, in terms of how I heard it on location. It's quite a challenge. The act has to be solitary. But I think you can share aspects of listening. It's a good question actually.

Does the fact that you are an experienced field recordist pose a problem for listening to sounds in a naive and unbiased way to extract their true meaning? Or does it, on the contrary, help you?
My experience helps me immeasurably to listen, as it would anyone. You can learn to listen. We can hear everything but when we take the time and the trouble to listen, then you hear things in a new way and get a deeper understanding of the subject. It's more interesting and involving.

I think listening in a naive and unbiased way is quite a strange way of phrasing it because you can listen to something in an unbiased way but the problem is we don't have ear lids. We can't shut off. We can go from listening to hearing ... and maybe that's what we do when we listen in a naive way.

In order to engage with something, when I'm giving a performance or presentation, I always set the context. That's important to me. I know some people might want the abstract effect of just listening to something, whether it's a refrigerator hum or a dawn chorus, but for me, people's sense of engagement is enhanced if they're not totally ignorant of the situation and have some idea of what they're listening to. If you want to get people to listen, you need to make the circumstances right for them to do that. It's quite a challenging thing to do. I like to try and describe something of my processes which might colour their sense of appreciation but it also might fire their imaginations in new ways which is the great thing about sound. I think it helps.

How do cultural differences shape our perception of animal song?
In Africa for example, they have a deeper, much greater understanding of sound than we do. The people I often get to work with in wild remote places have got ways of listening to the world which we are ignorant of. It's very exciting to learn from people who have a deep understanding of how their environment sounds. It's part of the way they get through life.

I was in Kenya a couple of months ago with a guide at Lake Bogoria, which is right on the equator. He took me to this amazing place called Mawe Moto, which translates to 'the place of the hot stones'. It's a place where there's fumes and sulphur and hot springs and he had all sorts of interesting stories about the two tribes that lived in this area. He spoke of how they battled, which is represented by the sound of steam issuing from these vents and the rumble of water underneath these stones. He said that if you listen carefully you can hear the story-tellers of these tribes recounting their history through the sounds of the bubbling water. It was fascinating to have that purity of content.

Another example, is with bird song in the rainforest where you hear everything but you see virtually nothing. The Baku people in the Congo are guided by the sounds in the forest, they navigate by the sound of the place and not by how it looks.

Through listening techniques like temporal resolution, and the possibility of playing sounds back at animals, the technical basis for a potential dialogue exists. Do you ever see it happening?
No. Simple as that, no. It's an interesting and amusing question. I think

we would be terrified if we could communicate with animals. It would all be over. I just can't imagine it and find it quite disturbing. My border collie has an amazing vocabulary of a hundred words, but just because he can respond doesn't mean he can understand or communicate with me.

The idea of temporal resolution is really interesting to me because we cannot approach the temporal resolution of virtually any animal I can think of. A classic example of that is the wren. The wren is a very common bird we describe as having a trill, and it can produce 64 separate notes in an eight-second burst of song. To us those notes appear as a trill. Whereas if you make a recording and slow it down, you can count the individual notes. You can count them but you can't understand them. In this blur of notes there's information about the bird's sexual status, its location within the habitat, its breeding status. All of that is conveyed in the fraction of a second, in a speed that we can't comprehend. And that's a relatively simple creature. Our temporal resolution can't approach the speed at which they can produce sound.

I think we can we get a sort of understanding, but I don't think we can get a full understanding. I'm sure people have made accurate definitions of what certain sounds mean but then there's a context as well.

What people like David Rothenberg are doing is interesting, but it's the frequency domain and the dynamics within which these animals exist that makes communicating with them difficult. Animals have been on the planet a lot longer than we have, and have had much longer to evolve sophisticated communication techniques. We might be able to transmit and receive information, but I think it's pushing it to say that we can communicate effectively with these animals. I think it's quite dangerous. Can you imagine what the world would be like if we knew what insects were saying? I mean, we can't even understand each other, so what hope have we got in trying to understand different species?

RUMBLE IN THE JUNGLE

Elephants prefer deep listening.

In March 2012, two herds of wild elephants gathered at the home of South African animal conservationist Lawrence Anthony. They had not been to visit the sanctuary keeper's house for over a year, but on the 2nd of March they made a 12-hour pilgrimage back to Anthony's home. It was the day that Anthony died of a heart attack.

It's well known that elephants mourn the death of their loved ones with a human-like solemnity. The herd will gather at the carcass and stroke it with their trunks and then proceed to collect branches to cover, and essentially bury the body. Up to two days is spent mourning in their funereal gathering and what's more fascinating is that elephants don't reserve this special ceremony for their own kind. Elephants have exhibited this behaviour towards other creatures also. There are many reports of elephants who display altruistic behaviours towards not only humans but other animals, assisting the wounded and protecting the vulnerable.

This amazing event surrounding Anthony's death opens up a whole host of questions about elephants and how they connect and communicate in this world. When Anthony opened Thula Thula, his conservation park, he had a reputation for being good with animals. In 1999 a troublesome group of elephants, who habitually broke out of their enclosures, was handed to him by local rangers. They believed Anthony could 'talk' with elephants, and true to his reputation, Anthony did indeed manage to keep the wild herd contained and safe within the sanctuary.

Anthony could reach these frustrated creatures who were violent and at risk of being poached and he ended up writing about his experiences in his best-selling book *The Elephant Whisperer*. Anthony was perhaps most famous for his rescue of animals from the Baghdad Zoo during the US attack in 2003, but it's his work with elephants that captured the hearts of the public. When the news spread about the elephants gathering at his home at the time of his death, it made headlines all over the world.

According to conservation charity Elephant Voices, elephants are the quintessential drama queens of the animal world. They like to make a big deal of everything and commonly experience emotions not usually associated with animals like joy, anger, silliness and indignation. They rush to any site of incident, quick to offer assistance and comfort and they let others know how they're feeling emotionally and in various other social situations. For example they call to warn others, to threaten, to announce needs and desires, to negotiate and discuss plans, coordinate group movement and assert dominance. Dr Joyce Poole, co-founder of Elephant Voices, now believes that elephants are capable of vocal learning and may even have different dialects among them, a very rare trait in the animal kingdom.

In terms of vocalisations, elephant calls are divided into two basic groups based on how the sounds are produced. Elephant calls seem to occur mostly during group coordination and mating periods and are mostly made by females. There are those that originate in the larynx and those made using the trunk. They've been heard rumbling, screaming, crying, barking, grunting and roaring, all part of the laryngeal group, and they can also create more tonal noises like a trumpet that are produced with the trunk.

Dr Tecumseh Fitch of Vienna University recently discovered that the unique rumbling noise made by African elephants is produced not by controlled muscle twitching, as in a cat's purr, but by the flow of air through the larynx. The vibration of their very large vocal folds is more like human speech or song, but it occurs at frequencies below 20 Hz known as infrasound, that we can't hear. Like humans, elephants can manipulate the structure of the sound by changing the shape of their mouth cavity and nasal passage to essentially filter the sound. Elephant Listening Project says that 'most elephant rumbles consist of a fundamental frequency between 5-30 Hz with audible harmonics or overtones.' The lowest call measured is 5 Hz for forest elephants and 14 Hz for savannah elephants. Not only low in frequency, these rumbles are powerful in terms of sound pressure, creating levels from 90 to 11 dB SPL, equivalent to heavy truck traffic or a construction site.

Elephants' infrasonic communication was first discovered by Katy

Payne in 1984 as she observed them in the Washington Park Zoo in Portland, Oregon. Rather than hearing the evidence, she could sense a vibration and suspected that the elephants were making sounds but they were below our hearing capabilities. Subsequent studies carried out in coordination with Joyce Pool, William Langbauer, Cynthia Moss, Russell Charif and Rowan Martin across Kenya, Namibia and Zimbabwe concluded that elephants use infrasound to communicate over long distances. When the temperature shifts at sunset, cool air gets trapped under a layer of warm air forming a kind of corridor that allows these low frequencies to travel much further. It is suspected that during this time of night, elephant sounds can travel over distances of up to 7 km through dense forest; not surprisingly, elephants are the most vocal during this time.

Another quirk of the elephant that went unnoticed until only a few years ago was their ability to mimic. Recent events have led us to believe that elephants, like bats, birds and marine mammals, are capable of vocal learning and mimicry, a feature used to strengthen and maintain unity and cohesion when the group separates. Dr Joyce Poole and her colleagues noticed one of the younger African savannah elephants making truck-like sounds that were unlike any of the calls they had heard from African elephants in the past. They realised that from her night compound, the elephant could hear the trucks three kilometres away on the highway.

Another incident occurred when researchers noticed that a 23-year -old African elephant Calimero, who had lived with two Asian elephants for 18 years, spoke in the Asian elephant 'language'. Unlike African elephants, Asian elephants typically use a chirping sound to communicate. Over time, Calimero had learned to chirp with his companions, almost to the point of excluding all the other sounds he once used. Researchers believe that this is the first kind of vocal learning they have ever seen in a non-primate land mammal. And it seems elephants can transfer this ability to other languages as well. In recent years an employee from the research staff at Everland amusement park in South Korea noticed that a 16-year-old Asian elephant named Kosik had learned to imitate human speech. Kosik was heard to repeat phrases that his handler used such as 'lie down' and 'good' and 'no'. He uses his trunk to help form the sounds that apparently even mimic the exact voice of the handler. It's believed that Kosik speaks clearly enough for others to understand, so much so that his special skill caught the attention of world media, with videos of Kosik popping up on YouTube.

Elephants are known to grieve, learn, play and empathise. They use tools, they cooperate and solve problems including figuring out how to escape elaborate bondage and enclosures. It is believed that elephants belong to the exclusive group of animals that exhibit self awareness, along with the magpie, great ape, and bottlenose dolphin. They are considered one of the world's most intelligent species with a relatively large brain that contains many similarities to the human brain. Certain elements of their neural systems can be compared to those of humans, including the size and complexity of the hippocampus, the area linked to emotion, memory and spatial awareness. In fact, their cerebrum temporal lobes, responsible for memory, are relatively much larger than those in human brains, giving rise to the old adage that 'an elephant never forgets'.

So perhaps when the elephants came to mourn their old friend Lawrence Anthony, it was because after all those years they had indeed come to understand him, not only through non-verbal communication, as he believed, but also possibly through verbal communication as well. But it still doesn't answer the question as to how they knew he had died. We still have a lot to learn about the way elephants understand this world, and perhaps the next.

---

http://www.birds.cornell.edu/brp/elephant/index.html (The Elephant Listening Project is a not-for-profit organization associated with the Bioacoustics Research Program (BRP) at The Cornell Lab of Ornithology in Ithaca, New York.)

http://www.elephantvoices.org

HOWLING BACK

Why do animals sing?

LANGUAGE AS A BASIS FOR MUSIC: IF WE COULD TALK WITH ANIMALS

The long-held belief that animals are incapable of language is beginning to crumble.

In 1995 George Johnson wrote an article in the *New York Times* that crystallised the animal language debate. It reported that there was no evidence that chimps could learn language and that they did not have the potential to create sentences of arbitrary length and complexity. Even Columbia University psychologist Dr Herbert Terrace, the man who in 1979 asked 'Can an Ape Create a Sentence?' concluded that the answer was a disappointing 'no'. Nim Chimpsky's trainer said that while a chimp might learn to associate a hand sign with an item of food, all it proves is simple conditioning, like that of Pavlov's dogs who learned to salivate at the sound of a bell. Barely 20 years ago this was the status of the great animal language debate.

The idea that animals are cognitive creatures began with Charles Darwin in 1872 with his book *The Expression of the Emotions in Man and Animals*. It was continued in 1926 when Jakob Johann von Uexküll developed his concept of Umwelt: the idea that your environment and experience is shaped by you and your perceptions. The idea was further developed in 1987 when Thomas Sebeok first coined the term 'zoosemiotics'. Since then we have overcome our outdated modes of thinking when it comes to animals and sentience. It took the scientific community until July 7th 2012, at the annual Francis Crick Memorial Conference, to accept that non-human animals have a thinking, feeling brain and consciousness leading to social and language systems.

Interestingly it is not fully accepted in the scientific community that there is such a thing as animal language. But of course this depends on how language is defined. In the 1960s, American linguist Charles Hockett outlined seven 'rules' of language. Hockett's rules suggested that to qualify as a language, a communication system has to have *semanticity*; an arbitrary basis; it has to be learned, not inherent; it must be comprised of units that can be used in many combinations to create many different meanings; it has to work on both a surface and semantic level; it has to be meta linguistic or have the ability to refer to the language itself; and finally it has to be able to communicate ideas about things that aren't in the vicinity either temporally or spatially. Looking at these rules it's no wonder theorists like Noam Chomsky regarded the concept of animal language as preposterous. Of course animals won't have the same systems as humans, of course they don't have the same vocal abilities, and obviously they don't all have the same level of intelligence.

In the 1990s, Dr Sue Savage-Rumbaugh wanted to prove to the world that animals, specifically primates, not only had a thinking, feeling brain, but that they had language and could learn ours as well. She spent years teaching Panbanisha the bonobo the English language and how to reciprocate using various systems like keyboards and symbol charts. The amazing abilities of Panbanisha made world news but after investigation, her claims were discredited and her life's work was snubbed by linguists and cognitive experts as irrational and ludicrous.

In Johnson's article, he suggested that some linguists and cognitive scientists felt that animal language experiments are motivated by more than scientific concerns. He went so far as to say that the experiments were motivated by a 'desire to knock people off their self-appointed thrones and champion the rights of downtrodden animals.' And indeed it's considered the ultimate faux pas in animal studies to anthropomorphise when interpreting the observations of animals. Conwy Lloyd Morgan went so far as to formulate a theorem known as Morgan's Canon to prevent anthropomorphic attitudes. The English psychologist stated that 'in no case should actions or behaviours be interpreted as the result of a superior psychic faculty, when it is possible to interpret them as a result of an inferior faculty.'

Today, it seems that researchers are taking a more balanced approach to the animal language debate. Audrey Parrish and Michael Beran from the Language Research Centre in the Georgia State University, believe that discussion about animal cognition in the past has been either completely inadequate, oversimplifying animals as 'stimulus response machines', or overly ambitious, imbuing them with impossibly complex intelligence. Professor of Musicology and Semiotics

Dario Martinelli thinks we should turn Morgan's Canon on its head. Instead of assuming an inferior faculty, we should in fact always suspect a superior one. Martinelli believes that Morgan's Canon inevitably leads to insufficient explanations and he proposes that rather than accepting the parsimonious notion that reality is actually simpler than it appears, why don't we instead assume that something much more complex is at play? A critical analysis of anthropomorphism by bioethics advocate Luisella Battaglia in 1997 concluded that emotional, 'human-like' manifestations that used to be considered inappropriate and even dangerous in scientific research can actually become valid clues to interpreting animal behaviours.

So can the scientific community find a happy place between parsimony and ideology? Dr Herb Terrace, when asked this question, recently answered with a resounding 'no.' He said that 'ideology has no place in science. Parsimony is the common goal'. He does however, agree that animals can think and not just in a concrete fashion but in abstract terms, about objects that aren't present. He doesn't agree though that they understand the difference between 'you' and 'me' and feels that that is the root of why they will never be able to have a real conversation. Surviving trends, public failures and peer opposition, the work of a small group of pioneers on the other side of the fence has gained some serious ground in the animal language debate. In 2011 Dr Sue Savage-Rumbaugh was listed as one of *TIME Magazine*'s 100 Most Influential People in the World and is now considered a bona fide trailblazer in the field of cognitive science. Her more recent efforts with Kanzi the bonobo lead us to believe that linguistic skills are not an exclusive human attribute and they reinforce Darwin's theory that 'nature does not move by leaps but through continuous gradations'. Dr Irene Pepperberg and her work with Alex the African grey parrot contributed considerably to field of language communication and comprehension abilities of animals. In a different approach Louis Herman, Roger Payne and David Rothenberg, instead of trying to teach animals to speak human, prefer to try and understand the animal language.

After spending many years studying whale song, Rothenberg made the discovery that these giants of the sea have a complex language and musical sensibilities, even if we don't understand what they are communicating. That failure to understand, however, is the ultimate obstacle in animal language research. Without a way to decode what they're saying, how will we ever understand their languages? Linguists call this riddle 'Gavagai', after the analytic philosopher Willard Van Orman Quine used the word to illustrate his Indeterminacy of Translation theory. In essence, he suggested that true translation is difficult to acquire because of the multitude of meanings and interpretations each word can have. A comprehensive understanding of language can only be achieved by studying behaviour in conjunction with language. You can imagine, then, the difficulty researchers have in finding appropriate opportunities to thoroughly observe animal language and behaviour in order to deduce even the most basic 'translations'.

There is someone who has overcome this hurdle, for one particular species at least. Dr Con Slobodchikoff of Northern Arizona University has spent over ten years researching the language of the prairie dog. He is the first person to effectively translate their vocalisations into a framework of meaningful utterances. They have a complex language of calls that can, for example, distinguish between the colour, species, size and shape of the intruder approaching. He was able to achieve these observations due to the particular environment in which prairie dogs inhabit, making observations, recordings, playbacks and further observations possible, therefore demonstrating the authenticity of his interpretation. Dr Slobodchikoff was able to prove, using the stringent rules of Hockett's theory, that prairie dogs do indeed have a language comparable to that of humans.

Other researchers want to move past Hockett's criteria altogether. They want to expand beyond the popular but inadequate concept of reinforcement as an analytical tool for animal language and cognition. The notion of reinforcement implies that animals are purely instinctive or, as Parrish and Beran so aptly described, simply 'stimulus response machines'. Research overwhelmingly shows that animals operate on a higher level than we previously assumed. Animals have integrative brain processes that mean they, like humans, form insights to challenges beyond mere survival and procreation. They too acquire rules, that take years to learn in order to perform tasks such as language, dialect, strategising, invention and highly evolved social functioning.

So what is the way forward for animal language studies? Dr Slobodchikoff believes that to progress, we have to do two things:

Change the paradigm of many linguists and biologists, from the one that says animals are incapable of language, to the one that says that it is very likely that many animals, if not most, have the capability of some form of language.

Set up creative experiments where we can associate the context with the signals produced in that context, so that we can build up a dictionary of meaning for each species.

Zoomusicologist Dario Martinelli envisages that as our knowledge of animals increases, so too will our understanding of ourselves, believing that this new perspective will put in motion some important changes. 'Who are WE?' Martinelli asks in his paper 'A (Personal) Introduction to Zoosemiotics'. 'Nowadays "we" corresponds to a larger number of individuals than just one century ago." Today, 'we' includes Australian aborigines, people with mental handicaps and even women. Martinelli hopes that 'we' will some day include many types of animals, not just the Great Apes which have been included in New Zealand's Animal Welfare Act. He states that the 'act is a serious threat to the humans-and-all-other-animals perceptive scheme' and he says that, 'this time, the draft is "Great Apes and all other animals". In a time not so far from today, it will be "Mammals and all other animals" and so forth.'

Reviewed by Audrey E Parrish and Michael J Beran, 'Thinking Animals: A Closed Case or an Open Debate?' in Frontiers in Psychology. 2012; 3: 250.
http://journal.frontiersin.org/article/10.3389/fpsyg.2012.00250/full

Dario Martinelli, OF Birds, Whales and Other Musicians; An Introduction to Zoomusicology, University of Helsinki

Douglas Broadfield, Michael Yuan, Kathy Schick, Nicholas Toth (Eds), The Human Brain Evolving: Paleoneurological Studies in Honor of Ralph L. Holloway

Dario Martinelli, A (Personal) Introduction to Zoosemiotics, http://www.zoosemiotics.helsinki.fi/intro.PDF

Con Slobodchikoff, 'Chasing Doctor Dolittle: Learning the Language of Animals', Northern Arizona University

WHY ANIMALS SING PART I: BEYOND SEXUAL SELECTION

Darwin's natural selection theory used to be the only official explanation as to why animals sing. The theory's monopoly over this issue is now being challenged.

Science aims to provide reason in a world where we have so many questions. So far it has given us reasons for almost everything that humans and animals do and have. Whether you call it biological determinism or evolution, most of the time it's hard to argue with the suggested theories. For example, the reason predatory animals have front facing eyes is to enhance visual resolution and better hunting, or the way water-loving creatures have body-cover that repels water and keeps the core dry.

But no one has yet to find a widely accepted reason for musical behaviours. Why do we make it? Why do we like it? Professor Daniel Levitin suggests music appeals to our brains in a two-part way. The first is our primitive lizard brain's desire for rhythmic expectation and the second is our higher functioning brain's delight when those expectations are smashed. Some suggest we like it because it lets us count without being aware we are counting – another apparently pleasurable activity for the human brain.

The most popular theory to date about the reason for music is based on Darwin's idea that we use it to attract a mate in the competitive game of sexual selection. But this theory is being increasingly questioned and is, like most other theories, inconclusive. Professor of Systematic Musicology and expert on the psychology of music Dr Richard Parncutt thinks his theory on motherese can 'account for just about every universal aspect of music, both the similarities with language and the differences'. Parncutt believes music began with the parental instinct to sing to their young and goes some way to explain its deep spiritual connection across all cultures. Some theorists are considering the idea that perhaps music was a protolanguage predating speech and that it has roots in animal song. So that begs the question, why then do animals sing?

We've discovered a whole array of animals, of all shapes and brain sizes, that produce musical vocalisations that we call song. Animals it seems, sing for many reasons, just like we do. They sing to announce dominance, mark territory, assist courtship, enhance social bonding and communicate over long distances.

Bird song is perhaps the easiest of all the animals' musical languages to identify. Researchers believe that the purpose of the 'dawn chorus' is to announce to all the single ladies that 'I have survived!' Birds often die in the night if they haven't fed well the previous day. So a strong and beautiful dawn repertoire carries meaning that the bird is a good finder of food and potential provider. Bird song also indicates the health and stamina of the suitor. It can be an efficient way to let all the other competitors know who is the best singer in the bunch and to move on if they can't step up. We're fairly certain that these are the reasons for bird song, mainly because birds predominantly sing during mating season and stop when the winter comes and also because females don't often perform these songs.

There are however, exceptions to this general observation. Female robins like to sing and some birds, including robins, still sing in winter even after mating season is long gone. Why are some species satisfied with a short call to attract a mate and display physical prowess when some sing unnecessarily long and complicated symphonies that employ the human traits of repetition and transformation? If we're talking evolution, what's the purpose of such a long, energy-draining performance?

If songs are to differentiate and delineate, why then do some neighbouring birds sing the same tune? Why do some birds have such a large repertoire? The brown thrasher is known to have up to 2,000 songs. Why do some birds sing duets?

Further study into the 'dawn chorus' has shown that it likely serves more than one purpose and not just in social signalling. The Austrian zoologist, Konrad Lorenz, famously wrote a diary containing observations of his 'pet' jackdaw. He noticed that after a nice meal when the bird was relaxed it would sit on a tree and sing. The bird would perform every song in its repertoire with the drama of a Shakespearean actor. Lorenz believed it was a form of play. This observation suggests that when birds are happy, fed, relaxed and without predators, they sing just because it feels good.

And birds aren't the only ones who baffle the evolutionary reasoning of animal song. Whale song is an area that has enjoyed much attention. It has been discovered that certain whales sing for reasons outside of basic biological function like mating, territorial announcements and location. Humpback whales sing long, complex and ever-evolving songs that serve a purpose we haven't yet determined.

Gibbons are famous for their dawn duets. A monogamous couple will sing a rehearsed duet as the sun breaks over the horizon. Researchers think they sing to solidify their bond and position within the group. We know that wolves howl, but did you know that they howl for all sorts of reasons? They howl when a baby is being born, they howl when they are lost, when they're reunited, after a successful hunt, and before a hunt but whatever the reason they will howl together; as it appears imperative to their social experience and expression.

So it seems humans aren't the only species whose preoccupation with music baffles science. We haven't found a suitable theory about why animals sing either. It's likely they, like us, sing for many reasons. So why then are we trying so hard to find out the reason for the universal appeal of music, song and rhythm? Perhaps there is no absolute reason. So far we've discovered that music is connected to several functions in the human brain like physical response and coordination, language and memory. Could music have the same effect on the animal brain? Even if it does, it still doesn't explain the emotional response to music, which is, in the end the most powerful and gripping element. How will we ever know if animals have the same emotional connection to music that humans do?

We have only just begun to understand the animal (human and non-human) brain. The riddle of music still confuses and polarises scientists today but in the end, we don't need science to tell us what music means. Our oldest primitive brains understood music; how it made us feel, how our bodies responded and how our minds were affected. If the most base, universal appreciation of music lies in our instinctive 'animal' brain, then chances are that animals feel the same way about it as we do.

Daniel Levitin, This is Your Brain on Music: Understanding a Human Obsession, Atlantic Books, 2007

Public Lecture by Professor Richard Parncutt, The origins of music: Grooming, flirting, playing, or babbling?' for the Centre for Music, Mind and Wellbeing at the University of Melbourne, 2011

WHY ANIMALS SING PART II: SOCIAL COHESION

Some animals may be singing to strengthen their group.

Animal song may have inspired generation upon generation of poets and composers, but it has mostly puzzled scientists. Even after centuries-worth of speculation, debate, data collection and research, there are still only two widely accepted theoretical approaches trying to explain why animals sing – and both are neither comprehensive enough to serve as an all-encompassing theoretical basis nor without their flaws. The first is guided by the understanding that animal music is part of the sexual selection processes. As Darwin himself put it: 'Musical tones and rhythm were used by the half-human progenitors of man, during the season of courtship, when animals of all kinds are excited by the strongest passions'. The second stands in open opposition to the adaptionist approach and, as laid out by William James in *The Principles of Psychology* (1890), regards music as 'a mere incidental peculiarity of the nervous system', and 'a pure incident of having a hearing organ', asserting that song isn't used to attract mates and has 'no zoological utility' whatsoever. The heated discussion between proponents of these two groups looks set to continue for many years to come. Over the past decades, meanwhile, a third set of concepts has established itself, which may come to explain at least a few of the loose ends and unanswered questions of the more established theories. Dealing with group dynamics and social cohesion, it has the additional benefit of bridging the gap between animal song and human concepts of music.

With humans, after all, it has never been questioned that music plays an essential role in social group processes. According to Aniruddh Patel, music, as a less contentious and yet more precise emotional language than words, has the unique capacity of balancing competition and cooperation between different community members. Juan Gualterio Roederer from the University of Alaska Fairbanks, a professor of physics with a love for music and a high proficiency in performing on the organ, was one of the first to attach the concept of social cohesion to music, emphasising 'the value of music as a means of transmitting information on emotional states and its effect in congregating and behaviorally equalizing masses of people.' While Roederer's theory has proven appealing to anthropologists, it has met with some criticism in the field of animal music. As Christine Cuskley points out, 'there isn't much evidence that music conveys information about things such as benefits an individual can provide', thereby questioning whether the costs of developing singing skills are truly outweighed by their benefits. It hasn't kept biologists following in Roederer's footsteps from searching for precisely that evidence. With important gains made over the last decade in particular, it seems as though they may have come thrillingly close to unravelling the mystery.

Synchronous calls
Björn Merker from the Royal University College of Music in Stockholm, Sweden, has been among the most productive and successful of the pack. In 2000, together with his colleagues Nils L Wallin and Steven Brown, he published a collection of essays, *The Origins of Music*, which sought to present an overview of the current state of research on the topic. It was his own contribution which would turn out to be most influential. Known as the theory of synchronous calls, Merker's approach is remarkable in two regards: Firstly, it presents convincing examples from the animal kingdom demonstrating the value of animal song for the group as a whole. And secondly, it argues that these songs may constitute the decisive link for the development of man from animals.

Merker's whole arguments rests on a strikingly simple thought. Looking at those features of music which are shared by all humans regardless of their cultural background or education should obviously yield the most basic, and therefore earliest, characteristics of music making. Curiously, there aren't all that many of these so-called human essentials in music. Merker decided to pursue the notion of measured music, i.e. music with an evenly-timed beat, and to search for it within the world of animals. He quickly found examples of synchronous pulses in 'lower animals', in fireflies, cicadas, crickets, crabs and frogs. In some cases, synchronous chorusing is used as a sort of default position: males will use it to set off their own signal against those of the others in a bid of reaping the benefits of what Merker calls 'a mechanism of temporal contrast in the perceptual system of the female which makes her turn to that out of several competing sounds which occurs just ahead of the others'. While this behaviour must be considered competitive rather than collaborative, Merker offers various examples of the social kind as well, with cooperative synchronous chorusing mostly being used to increase the geographic reach of a group of males looking to attract potential mates. Synchronous emissions of the same signal create an amplitude summation which can spread the group's song to regions uncoordinated individual voices could never reach: 'Well synchronized groups of males would thus tend to attract females to themselves at the expense of less well synchronized groups.'

Synchronous calls are present in 'higher animals' as well. Merker singles out chimpanzees in particular, who use synchronised calls for a variety of reasons and situations. Males will use synchronous calls to attract migrating females, for example. Or they may use them for the so-called 'carnival display': If a chimpanzee discovers a valuable food resource, such as a tree carrying ripe fruit, he will send out a call to the other members of the group. As soon as a new chimpanzee arrives at the location, he will immediately join in, thereby amplifying the signal and attracting even more friendly males – as well as females from outside territories. Thus, social singing serves three purposes: The sharing of resources among group members, which increases its chances of survival; the attraction of female chimpanzees to the territory, which increases the chances of progeny; as well as a sort of control on valuable calls and non-valuable ones: Should there be a false alarm (i.e. if the tree turned out to carry no edible fruit), new chimpanzees arriving on the scene may retaliate. To Merker, these aspects are so beneficial that they can come to account for the split between early monkeys not capable of synchronous calls – and those who would eventually come to develop into human-strands.

Territorial advertisements

Merker's findings are intriguingly sandwiched in between the aforementioned Darwinian selection and another complex of social cohesion theories under the moniker of 'territorial advertisements'. In

'Did Neanderthals and other early humans sing? Seeking the biological roots of music in the territorial advertisements of primates, lions, hyenas, and wolves', Edward H Hagen and Peter Hammerstein pick up Christine Cucksley's question – 'What could a song, especially one without words, tell us that we do not already know?' – as a point of departure. Their ideas are based on surprising insights from the concept of territoriality, i.e. 'the defence of an area to exclude other animals, using physical force, threat, or advertisement'.

If one simplifies this down to a situation between an owner of a territory and an intruder, then a surprising fact emerges: The owner is usually capable of defending his claims even if the intruder is marginally stronger than him. Although this may seem paradoxical, Hagen and Hammerstein convincingly explain it as the animal equivalent of a mutually beneficial convention: 'In the U.S. if two automobiles reach a four-way intersection at the same time, the convention is the auto on the right has the right-of-way to proceed through the intersection first. Although each driver has an incentive to proceed first, each has bigger incentive to avoid a collision. By agreeing on a convention, any convention, for the right-of-way, all drivers avoid the large cost of a collision, albeit at the smaller cost of ceding right-of-way half the time (on average). Similarly, by agreeing on the convention that intruders always retreat, animals avoid the cost of fighting, albeit at the cost of ceding territory when in the intruder role (but keeping it when in the owner role).'

With this in mind, it obviously makes sense for the current resident of a territory to defend his claims by announcing his ownership. This is possible by scent marks – 'strong-smelling urine, faeces, or secretions from special glands are rubbed or deposited where intruders are likely to encounter them' – or by means of acoustic signals. Hagen and Hammerstein offer various fascinating examples of these, from the drumming of the bannertail kangaroo rat, which mostly lingers on or in a huge storage mound, defending its contents of valuable seeds by drumming its feet, to the song of birds. Wolf howling, too, is a fascinating phenomenon, especially since it serves more than just territorial defence strategies. The individual and collective howl is not just a statement of presence but reveals vital information about the size of the pack, the presence of young as well as pack composition. As Charles Burr points out in the liner notes to *The Language and Music of the Wolves*, 'each wolf has a position in the pack, somewhere below the Alpha wolf, the leader. Wolves live together peacefully by mutual understanding dependent on their signals to each other – their body language, the carriage of their tails, their facial expressions and most particularly their barks and howls. (...) The wolf is a wanderer. In order to re-establish contact with his other pack members, the individual wolf raises his voice. Wolves (...) cry out to be heard, so that others may tell who they are, and where. And howl back.'

Aniruddh Patel, Music, Biological Evolution, and the Brain, http://cnx.org/contents/8aa30a5e-9484-4d55-87e3-d53c1cbfb8b2@1.4:8

Christine Cuskley, Evolution of Cognition

Bjorn Merker, 'The Birth of Music in Synchronous Chorusing at the Chimpanzee-Hominid Split', http://www.escom.org/proceedings/ICMPC2000/Wed/Merker.htm

The Language and Music of the Wolves

Edward H. Hagen and Peter Hammerstein, 'Did Neanderthals and other early humans sing? Seeking the biological roots of music in the territorial advertisements of primates, lions, hyenas, and wolves' in Musicae Scientiae 2009 13: 291

EVERYTHING IS WRONG: BERNIE KRAUSE'S CONCEPT OF BIOPHONY

*If Bernie Krause's theories are true, then animal song is part of a
far more complex and all-encompassing sound world.*

There was a time when Bernie Krause listened to the world like anyone
else. He cherished the song of a robin in the park, trembled at the roar
of a lion at the zoo and was overwhelmed by the awe-inspiring volume
of a tree frog's mating call. After experiencing an epiphany on his first
wildlife recording session with headphones, however, Krause traded
in an extremely successful career as an audio engineer, synthesizer-
specialist and songwriter with renowned duo Beaver & Krause for a
calling as a field recordist. He set about committing the world around
him to tape, eventually creating the perhaps most expansive private
collection of animal sounds, today clocking in at 4,500 hours of material
by 15,000 different species. It would take several years for Krause to
realise that this catalogue-approach was fundamentally flawed and only
capable of reflecting a tiny fraction of the vast sea of sonic information
contained within natural soundscapes. What he was doing, he realised,
was 'like taking apart a Beethoven symphony and just examining the
strings lines or the horn line. And you listen to it and you say "Well,
that's really nice"'. But it didn't tell you what the piece was about. For
the next decades, he would make it his life's mission to uncover the
meaning behind the sounds.

Decontextualisation

Until the moment when he realised his mistake, the prevailing take on field
recording was an approach he would later refer to as the 'decontextualised
single-species model'. Krause traces back the origins of this model to
the achievements of Ludwig Koch, a German-born sound recorder who
had escaped to England during the Second World War and accepted a
position at the BBC, as well as a team of ornithologists from Cornell
University. Their pioneering efforts in collecting the sounds of birds in
Great Britain (Koch) and at a 'gator-infested Georgia fen' (the Cornell
team) would dominate public debate on the topic until the present day
and define field recording as being about 'the idea of life lists – finding

and identifying single species of birds and mammals and, more recently, frogs and insects'.

It would take until the early 1980s, when Krause was in Africa on a mission to record the audio part of an exhibition about a waterhole and the animal world gathered around it, before he started to feel otherwise. After setting up his gear, he was instantly struck by the richness and magnificence of the creature voices and decided to tape a far longer passage than he'd originally planned. Depleted by the hard work of the day, he had crept into his sleeping bag and was listening to the soundscape. It was then that he realised something seminal: 'It was in this semi-floating state – that transition between the blissful suspension of awareness and the depths of the total unconsciousness – that I first encountered the transparent weave of creature voices not only as a choir, but as a cohesive sonic event. No longer a cacophony, it became a partitioned collection of vocal organisms – a highly orchestrated acoustic arrangement of insects, spotted hyenas, eagle-owls, African wood-owls, elephants, tree hyrax, distant lions, and several knots of tree frogs and toads. Every distant voice seemed to fit within its own acoustic bandwidth – each one so carefully placed that it reminded me of Mozart's elegantly structured Symphony no. 41 in C Major, K. 551.'

Krause realised that the separation of animal song into individual voices did not present a sensible reflection of the world around us, because animals simply did not experience the world this way. Their acoustic habitat was marked instead by a deep, seemingly impenetrable constant togetherness of events, a simultaneity of myriads of different signals, the ebbing and swelling of sound from quiet moments into glorious, anthemic dawn and dusk choruses. Taking into account the importance of their calls for protecting their territory and attracting suitable mates, every lost emission was either a threat to their own life or the survival of their species. Which left only one possible conclusion for Krause: Animals had to find ways of cutting through this wall of sound and they did this by claiming their own frequency in the sonic spectrum. And so, the lower ends are usually taken by mammals, from the subsonic utterances of giraffes, elephants and hippos to the sounds of monkeys and cats. Further up in the spectrum, one finds different species of birds, which have arguably taken the art of song to an unprecedented degree of refinement. The highest frequencies, finally, are secured by insects and the ultra-sonics of bats. Together, they form what Krause termed the 'Great Animal Orchestra', a constantly shape-shifting constellation of individual voices in motion, and he termed their symphonic soundscape a 'biophony' – all of the 'sounds originating from nonhuman, nondomestic biological sources'. Krause's thesis was instantly plausible and had far-reaching consequences not just for fellow musicians, but also in the realm of ecology. Which made field recording more political than it had ever been before.

Acoustic evolution

Krause's theory certainly had the advantage of not just presenting the most detailed explanation model for the acoustic environments surrounding us today, but actually offering potential insights into the evolution of species on the basis of sound. Before there was animal life on the planet, after all, the only noises were those springing from 'nonbiological subcategories such as wind, water, earth movement, and rain'. When life started to emerge in the giant oceans around 600 million years ago, the first organisms to populate this 'geophony' still had essentially unlimited bandwidth to choose from. Slowly but surely, however, things became more complicated: 'At first, when their numbers were relatively small, acoustically sensitive organisms merely needed to filter out the geophonic background in order to perceive other sound-producing organisms within their habitats. As the number of species increased and became more complex, they had to be able to hear and process the particular sounds that were relevant to their well-being. Over the course of many glacial periods, especially the recent ones, the total number of creatures multiplied exponentially – species filling available biological niches. Complex habitats arose that supported robust varieties of life-forms whose behavior and survival – both individually and collectively – were determined to a large extent not only by visual, olfactory, and tactile cues but by sound.' What we're hearing when we're listening to animal sounds, therefore, is anything but the spur of the moment, but the product of millions of years of adaptation and transformation, of individual changes within a huge sound body filled to the brim with information, in which every part is related in intricate ways.

If Krause is correct, then our ancestors were capable of extracting this information and using it to their advantage – either by using it as a sonic compass for navigation, for recognising potential dangers (such as the proximity of predators or changing weather conditions) or by imitating animal calls as a bait or as a means of appeasement. As Krause puts it: 'The ability to correctly interpret the cues inherent in the biophony was as central to our survival as the cues we received from our other senses.' Our ancestors knew that the biophony of a particular location can tell us a great deal about factors such as which species populated the area, how many of them there were and where, roughly, they could be found. In short: It told them a lot about the health of that location. Some of the instincts from these times are still intact. Chris Watson has reported on the powerful sound of places in the Kielder Forest in Northumberland, where he would sense inexplicable, bad feelings about some of the places. And yet, only a few hundred metres away, he could set up his gear and feel perfectly happy. Intrigued by this phenomenon, Watson would ask guides on all of his trips around the globe whether they knew about similar places. Astoundingly, all of them did. The recurring pattern seemed to be that the 'evil sounding' spaces were those devoid of animal noises, while those full of life and sound tended to come across as far more agreeable.

For one of his projects, Krause gained permission to record at a forest management area called Lincoln Meadow in the Sierra Nevada mountains, both before and after a lumber company had performed so-called selective loggings. The result stunned him. In 1988, shortly before the company set to work, he recorded the biophony in the last hours of daylight. The outcome was a spectrogram with a remarkable density throughout all frequency bands, as could be expected for a habitat replete with the most diverse animal life. In 1989, he returned to the meadow after the operation had been completed for a second session under the exact same conditions and at the exact same time. In keeping with what had been promised by the logging company, the place still looked as though it was teeming with life – 'I was delighted to see that little seemed to have changed', as Krause remarked. Back in the studio and after a look at his spectrogram, he had to revise that impression: 'Gone was the thriving density and diversity of birds. Gone, too, was the overall richness that had been present the year before. The only prominent sounds were the stream and hammering of a Williamson's sapsucker.' The ear, then, turned out to be capable of detecting the true state of the habitat much more precisely and truthfully than the eye ever could.

Krause's work has done a lot in terms of explaining why human acoustic presence – the anthrophony – can be harmful to habitats. The volume of passing cars can result in birds living close to a highway no longer being able to hear and locate each other and thus, in a decrease of their population. Likewise, low-flying planes can disrupt the synchronous calls of frogs or cicadas. Once that synchronicity has been broken, individual voices become audible, making them vulnerable to predator attacks. What's more, it casts doubt on the so-called progress that has been made so far in trying to understand animal song. If that song is always the result of a far larger composition and integrated into a long development starting at the very earliest stages of life, then this strongly suggests our concept of producing sound is most likely closely integrated with that of all other species on the planet. And our analysis of animal sounds and song would never be complete without taking this into account.

Gino Robair, Going Wild with Bernie Krause http://www.emusician.com/artists/1333/going-wild-with-bernie-krause/37582

Bernie Krause, The Great Animal Orchestra, Back Bay Books 2013

CHARMING THE CHIMP

Some artists are attempting to enter into a dialogue with animals.

AN IDEAL AUDIENCE: THE WORK OF MAREK BRANDT & DAVID TEIE

Marek Brandt and David Teie are composing music for insects,
gnats, frogs and monkeys. What happens when animals turn
into listeners?

Marek Brandt doesn't even own a pet. And yet, over the past few years,
he has shared some of the most magical moments of his life as an artist
with animals. Once, he went looking for elks in the Swedish forests only
to end up giving a concert for gnats. On another occasion, he waited for
hours to play recordings of the mating song of frogs from his hometown,
Leipzig, to their remote relatives in a French pond. On Gibraltar, Brandt
drove a borrowed car up the island's spectacularly barren chalk cliffs;
once he'd arrived at the top, he heard a strange sound to his left – and
a tiny monkey's arm squeezed its way inside the vehicle. At the Polish-
German border, he and a group of befriended musicians imitated
howling wolves, much to the bewilderment of the passers-by.

By all accounts, the German sound and visual artist, photographer
and designer has garnered plenty of success with his events: Despite
its inherently experimental nature, his work has been featured on
German cultural TV station 3SAT and local radio, from underground
publications to nation-wide newspapers, at small-scale events and major
festivals. And yet, although not a single animal has ever been harmed
during any of his events and response has generally been positive, there
has been plenty of criticism as well – and not all of it has merely been
verbal. In February of 2007, Brandt presented his installation 'Mouse
House' to the public. It consisted of a small shelter inside a cage made of
plexi glass, a miniature-scale reconstruction of Barcelona's Formula 1
race course, a feeding dish, speakers emitting a sound art composition
as well as the installation's inhabitants: two African harvest mice. It
seemed a perfectly innocent set-up. Still, on the last day of exhibition,
two activists associated with the vegan scene misappropriated the rodents
as well as their temporary home, destroying the artwork in the process.
For such a small endeavour, it seemed a somewhat overblown statement
– what on earth made these people so angry about Brandt's work?

Emotions & symbols

Brandt was initially shocked by these reactions. Today, he can see why his 'music for animals' should garner such heavy emotional responses. After all, the notion of entertaining animals and writing pieces for them is merely a superficial layer on top of a far more complex and, as the artist puts it, symbolically charged project – one he's been labouring over for roughly a decade. In the early stages, Brandt intended to transfer the heart sounds of sheep to the meat counter of a supermarket. The idea would have been hard to realise even under the best conditions, but was, in the end, stopped dead in its tracks by the outbreak of foot-and-mouth-disease in the UK, which meant farms became inaccessible to the public. This setback didn't deter him from his ideas, however, and in 2004, he received a travel grant from Saxony's ministry for science and the arts specifically for his animal-music concept.

What followed were six months of travelling and performances. Brandt began visiting the Max-Planck-Institute to gather information for his pieces: 'I'll always begin with researching the species. A question I'll ask myself is whether there are any insights about which sounds and instruments are accepted by particular animals. If you're playing in front of ants, for example, you shouldn't use any extremely deep bass frequencies, since they could potentially endanger the hive – even the tiniest of vibrations are transmitted through the earth and the ants will direct their senses towards them. Bulls, on the other hand will, despite your best intentions towards them, run away from noise and reflections from metallic objects. When it came to gnats, in lack of prior experience, I had to come up with a solution myself. In the end, I opted for a combination of flute and electronics.' Does it work? Not always, as Brandt freely admits – some of the recipients simply decided to swim away or sting him. Then again, there were moments of success, which made up for all the disappointments. In Leipzig, he flew a balloon with a speaker close to a group of crows in a tree, which seemed intrigued. And back in Gibraltar, where that tiny monkey had invited himself into his car, the reaction was almost human: The animals listened with seeming interest for 15 minutes, then went off to do other things instead.

Needless to say, some people consider Brandt somewhat of a Gyro

Gearloose. Which is wrong in more than just one way. For one, he is not a scientist. The biological and neurological aspects are merely departure points for him, never a goal in themselves. And, as much as he is honestly interested in his audience's response, there is a variety of adjacent topics which play into his music for animals: 'The original idea was to consider and reconsider our perception of animals in a society, which is increasingly losing touch with nature – at least in Western European cities and metropolitan areas. It was also important to me to examine their historical importance as symbols both in the world of the arts and in society. One shouldn't forget, after all, that many species have become an integral part in flags and religions as symbols and heraldic animals. And yet, how many of these identifiable codes and symbols are still in place in our contemporary, hyper-fast society, which mainly regards animals as productive livestock? What made the concept interesting was that it provided me with new insights even in case of a "failure" and to include the imponderable into my work. I am an artist and a musician, so my task is different from that of a scientist.'

Animal specificity
More than 4,000 miles across the Atlantic, David Teie is working on a strikingly similar idea. Although most people will know him for his remarkable cello technique, his job as an instructor at Maryland University or his string arrangements for the side-project of Metallica's Jason Newsted, Echobrain, Teie's interest has always extended far beyond the familiar conventions of music. One of his projects is a multi-angle recording and headphone-playback-technique which could revolutionise recorded music. He has also added to the debate about the origins of music, arguing that human music is the way it is because it was written by humans for humans. As it turned out, he had to turn to animal song to prove his hypothesis: 'I began by trying to come up with a theory on why music affects our emotions. I separated music into indivisible components and asked the question of each: why would this element induce an emotional response? After a few years of research and conjecture I had gathered plausible answers for each element and thought I had something of a complete theory of the origins and affective

processes of music. Any good theory is testable, and one of the tests of the theory would be that I should be able to take the principles of human music and apply them to other species. I studied the emotions of animals to see what kind of sonic triggers to their emotions I could find. If my theories were correct, I should be able to create species-specific music for animals.' It was a perfectly plausible consideration. At the same time, it would provide him with his biggest challenge as a composer.

Teie's first step was to send his book to renowned Stanford neuroscientist and author Robert Sapolsky. Sapolsky seemed an obvious choice: For more than a quarter century, he'd observed and analysed the behaviour of a single group of baboons in Africa and gathered extensive knowledge about their behaviour. Unfortunately, although he seemed interested enough, the baboons in questions were, as Teie put it, 'nasty and not good subjects for my kind of study'. However, Sapolsky did provide Teie with another contact, Charles Snowdon, Professor of Psychology and Zoology at the University of Wisconsin-Madison. Since Snowdon wasn't prepared to jump right into the project, it took a little convincing to get him on board, as Teie remembers: 'One time he sent me two new recordings of two recently discovered calls of the cotton-top tamarin monkeys. The calls were named similarly: the SL multi and the SL trill. I analysed them and wrote back to him that I thought they did not belong in the same category based on a musical analysis of the calls. One sounded like a threat and the other sounded like an in-group vocalisation. As it turns out I was right even though he hadn't told me the context of the calls. He liked the idea that based on a musical analysis I was able to tell him generally what the context and meaning of these calls were. After that he decided to pursue this study with me.'

Tuning into the monkeys' system

Teie's approach was simple: Study the vocal expressions of animals to arrive at a sonic syntax tailored directly at them. An article for the *Washington City Paper* correctly established that it was wrong to conclude that monkeys did not like any kind of music just because they preferred silence to human music – or that, as Joshua McDermott of New York University claimed, music 'is only going to obscure sounds that are meaningful to them, like the sound of an approaching predator or the call of a nearby monkey.' Instead, Teie was convinced that one had merely to tune into their sensory system to leave an impression. He used vocalisations of the cotton-top tamarin found in an online library and slowed down the recordings by a factor of anything between four and ten to adjust them to the human vocal range and to transcribe them. To Teie, it was clear from the outset that the two determining factors for writing species-specific music were the vocal range as well as the respiratory rate of the adult female – the latter of which was in tune with the notion that the heartbeat and breathing rate experienced by the foetus in the womb are of seminal importance to their appreciation of sound. After hours and hours of work, he finished two different compositions for his new audience: 'exciting monkey music' and 'monkey lullabies', both of which sounded equally disturbing to human ears. He took them with him to Madison to observe their reaction. Before the tests began for real, meanwhile, he couldn't resist the temptation to try out his work in practise: 'In one visit I tried a simple version of what would eventually be my music for tamarin monkeys and it was very interesting to me that it was immediately quite successful. I whistled a tune that was something of an improvised set of variations of affectionate tamarin vocalisations. The monkeys stayed in place (very unusual for this species that is constantly hopping from place to place as a defence against potential predators) and stared at me. It was a stunning and magical moment for me.'

It wouldn't be the last and only magical moment. When Snowdon and his team played Teie's compositions to the monkeys, they reacted just the way they had anticipated and hoped: The exciting music sent the monkeys into a state of heightened activity and induced them to mark their territory. The relaxing piece sent them into a state of tranquillity. In their paper, published in the *Biology Letters* later that year, the results still sounded fairly dry. The bottom line, however, was remarkably in your face and as spectacular for zoologists and musicologists as for the average visitor of the local zoo: 'Music composed for tamarins had a much greater effect on tamarin behaviour than music composed for humans.' The implications of this statement were so far-reaching, that they seemed to mark a watershed: Were humans at the brink of truly communicating with animals? Were animals at the brink of appreciating art? The questions got increasingly spectacular and speculative and suddenly both Snowdon and Teie found themselves at the centre of media attention, conducting dozens of interviews a day and receiving more attention within a few weeks than over the course of their entire professional careers.

Yet more questions

In truth, as with any meaningful scientific study, Teie and Snowdon's may have answered one or two questions, but merely opened up even more in doing so. One of them was related to the notion of 'form' in animal music. To arrive at his monkey pieces, after all, Teie had mainly worked with sound and pitch, in a bid to reproduce and rework certain characteristics of the tamarin language. His compositional procedures, meanwhile, had been entirely human and consisted of repeating the sounds or pitching them up and down certain intervals. Shouldn't monkey music take monkey syntax into consideration as well? 'The question of the adaptation of forms to a given species is one that I have yet to consider, although it is a very interesting and valid consideration', Teie admits, 'For now I am adapting our forms, ABA for example, to the music for animals. It does seem fitting that the pace of variation in our music is calibrated to human perceptions. Brahms, for example, often changes patterns every 10 to 30 seconds or so. Surely this is accurately attuned to our own pattern sensibilities. When Tchaikovsky seems to go on a bit too long with a pattern, such as the French army retreat passage in the 1812 overture, many think it gets a bit tedious. Perhaps animals who process sounds 10 times faster than we do and whose vocalisations need to be slowed down 10 times in order for them to be heard in our own vocal range would find pattern changes to be optimal every 1 to 3 seconds! This is a very discouraging thought to me as a composer of species-specific music, considering the amount of time it takes to research, write, and record a piece. I imagine spending a week producing 12 seconds of music.'

Other questions directly affected the process of actually composing the music: In which way was there still a sense of satisfaction when writing in an entirely foreign musical tongue, the results of which would never be open for his own appreciation? And what constitutes quality in an environment when the only gauge of a track's effectiveness was how soothing or stimulating it was? Teie was able to resolve at least part of the dilemma when taking on his second big animal music project: Music for cats. Compared to the tamarin experiments, the endeavour had two decisive advantages: With many millions of cat owners, for one, Teie was able to test its efficacy on a far wider audience. And should his music truly work in improving the moods of felines all around the world, he finally stood at least a chance of financial remuneration for many years of hard, unpaid work.

In terms of artistry, on the other hand, the challenge was even bigger than the monkey music. Much of cats' hearing is situated in the ultrasonic range, meaning that desperately little is even audible to humans. To keep things interesting for himself, he decided to compose two pieces at the same time: 'I found it difficult to continue writing music for cats unless I included that portion of the songs that is intended for the cat owners as a part of my inspiration. There is a frequency range within the hearing range of cats that is below their vocal range that I presume is not subject to their appreciation of music. We humans, for example, hear sounds above and below the frequency range of our own voices but we don't recognise pitches in these ranges. If this is true, then the range of the male human voice is not of interest to cats in terms of the same kind of emotional response they would get from a vocal communication from one of their own species. I use this range then to write music for the cat owners. The music in this stratus may be something like background noise to the cats. It is like writing music with traffic noise included. It should be ignorable in the background for the cats and intended for the Homo sapiens in the room. I have found that it's easier for me to write music when I consider this layer of music to be the pathway through which I communicate as a composer.'

Surprisingly for someone as immersed in writing music for animals as Teie, a two-way communication is not even remotely on his radar.

Miracles, as he has noted, mean nothing in the world of 'savage beasts' and music is unlikely to calm a raging manatee. And yet, what he does believe in is furthering a deeper understanding of animals and, by means of comparison, ourselves. It is a remarkable similarity with the work of Marek Brandt. For Brandt, playing his music for animals, we reflect upon ourselves and our own perception of sound, because animals represent possibly the ideal audience: One which may be listening to their very first piece of music ever (or, as in the case of the one-day-fly, their only piece ever), without any preconceived notions, expectations or suggestions about how to react. It is the fulfilment of an age-old dream of every true music aficionado: To be able to return to that state of perfect naiveté and bliss, which filled us when being exposed to music as a child. To be able to get in touch with this wonder again, the world of animals offers sound artists and composers a unique opportunity, according to Brandt: 'Whenever I'm playing for animals, I get the feeling of being able to work much more open and experimental than with humans. It is a lot harder to predict the response of the public. But that is precisely what makes it so exciting: To constantly re-invent myself for the animals.'

Charles T Snowdon and David Teie, Affective responses in tamarins elicited by species-specific music, Department of Psychology, University of Wisconsin, Madison

http://www.musicforcats.com

THE GRASSHOPPER
AT TWILIGHT

But … is it music?

THE GRASSHOPPER AT TWILIGHT

A long night spent discussing all aspects of the animal music debate
– and a possible conclusion.

You can't help but feel that Slavek Kwi is always busy. When he's not watching over his two little boys, leading experimental workshops for children with learning disorders or travelling the world in search of new sounds, he can mostly be found at his home studio in Ireland, organising field recordings into compositions or into immersive 'cinema for the ears' radio plays. So naturally my request for a conversation about his sonic work in the Amazon rain forest is met with friendly agreement – but a note that making an appointment may be an issue. This hardly registers as a problem, since it gives me time to prepare and delve deeper into his personal history, which almost reads like a novel itself: After leaving former Czechoslovakia in 1986, 'For a variety of reasons', as he puts it, Kwi spent two years being lost on the streets of Germany and the Netherlands, before finally settling in Belgium, where he would stay for almost one and a half decades, becoming a part of the burgeoning tape scene and developing his now almost obsessive interest in sound and the art of arranging it. The move into music must have been a surprise for Kwi as well, who started out painting as a boy, committing the world around him to the canvas with oil colours. What he eventually discovered, however, is that these different modes of expression are really all connected by a single underlying theme: His connection with nature, which has gone from a sense of otherness (with humans standing outside of it, observing) to a close and intimate relationship (with humans fully integrated with it, sharing).

When I contacted Slavek, I told him I was interested in his recordings of the boto river dolphins of the Amazon. Which wasn't untrue. After all, his LP *Boto [Encantado]* on the Belgian label ini.itu had been a personal favourite of mine, an album so packed with sonic wonder that, at one point, I was occasionally under the impression that I had been on location as well and shared in the experience. It is also exemplary of his work in general in that it seems to occupy a grey area in between the poles of pure and processed recordings – a division Kwi uses quite

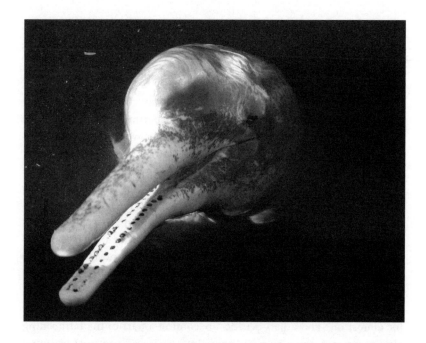

strictly on his website, but which is less down to a tangible difference in kind and more to his tendency 'to create a kind of confusion.' And yet, of course, I have also come here for answers to the big question of what, precisely, constitutes music from his point of view and how are animals participating in it? Since Kwi's oeuvre seems to be placed exactly at the cusp between field recordings, transformative sound art and more conventional definitions, he seemed to be the perfect sparring partner in this respect. As if to confirm my suspicions, the first thing he tells me after I've related some of the backgrounds of the project is: 'I love natural sounds, particularly animal calls, I find them absolutely fascinating.' His releases certainly reveal as much. Over the years, Kwi has collected the sounds of insects in the Czech Republic, of electric fish in the Amazon, jungle habitats in Southern Mexico and a variety of ultrasound-emitting species from around the world. And yet, the thing which probably intrigued me most were the liner notes to *Boto [Encantado]*, which Kwi had printed in mirror writing so as to offer the listener the opportunity not to be distracted by the technical detail of these recordings. But does knowing less really provide for a solid basis to discuss the most relevant aspects of animal music? It's time to find out.

'How did the *Boto* project come about?' I want to know.

'The idea to create a piece based on sounds of the boto came after my first trip to the Amazon: a remote reserve Xixuaú Xiparina on Rio Jauaperi off Rio Negro.' According to Slavek, 'from my scant research about the location, I knew the dolphins would be there. Though I was hoping to record them, I was not solely focused on dolphins – I wished to experience the place in all its possibilities, and simply record sounds within the habitat – both underwater and above. The concept: I was interested in the influence of information on the reception of the sounds, exploring perception from the perspective of both art and science, a combination of sound-art as purely an aesthetic experience together with scientific cognitive information ...' smilingly, he corrects himself, '... in this case pseudo-scientific as I am simply creating a model of thinking-geometry. The audio-material with river dolphins seemed to me especially suitable for this project: the variety of sounds generated by the boto, the acoustic richness of habitat, and the mythical context, all influenced my choice. 'Boto' is the local name for Amazonian Pink Dolphins, *Inia geoffrensis*. It is believed that from time to time boto come out of water as handsome men to seduce women. When a woman has difficulty identifying the father of her child, she blames the boto ...'

'The liner notes mention you recorded at two completely different sites, the Rio Negro and the Mamori. What are they like?'

'The Rio Negro has so called "black water", which is actually a tea-like colour and is acidic. In general, there appeared to be fewer subaquatic insects – though some sounds are astonishing, almost electronic-like – and frogs, different species of soniferous fish – it seemed a little bit quieter than the white water. One interesting fish song from Xixuaú Xiparina sounds like knocking on a wooden board at various speeds – this rather large fish, according to locals, has inside the skull a small cavity with two marble-like bones. The fish knocks together these

PHOTO: Asier Gogortza

balls, generating loud clicking sounds which you can hear even above water. Mamori has "white water", which is a yellowish milky colour and alkaloid, rich in sediments. In general, there is an abundance of insects and frogs, various soniferous and electric fish, crustaceans; sonic life seemed very proliferous. Aside from the "boto", there is another Amazon dolphin, "tucuxi". Tucuxi is a local name for a smaller estuarine dolphin (*Sotalia fluviatilis*). Unlike boto, who are purely river dolphins, tucuxi sometimes venture into the open ocean. Mamori is more populated than Xixuaú Xiparina; you can often hear small boat-engines. In Xixuaú Xiparina it is rare to hear such interference.'

'How did you reach these sites?'

'To get to Xixuaú Xiparina takes about 36 hours on a rather noisy double-decker boat from Manaus, up the stream of Rio Negro. We slept in suspended hammocks, myself and my wife Helen. During the night there were myriads of flying insects – mainly moths, sometimes very big, and startling colours – attracted by the light. The reserve is in the region of Rio Jauaperi off Rio Negro. There is a village with a small indigenous population, which is where the Amazonia Association is run from. The rainforest around is annually flooded – water rises and descends about 11 m – however, the surrounding forests seem to be still partly underwater all the time. The Amazonia Association organised a guide for us. The majority of the time we travelled in between the trees in a canoe. Flooded rainforest is incredibly beautiful; it has a spacious and resonant quality. Hiking through dense rainforest was rather limited; from those hikes I have long duration sequences when I left a recorder for several hours without my presence.'

I remember the pictures I've seen of the Xixuaú Xiparina region, an uninterrupted, vast stretch of rainforest - 600,000 hectares in total - almost entirely left intact. The Xixuaú Reserve consists of little more than a few straw huts by the river, home to 22 families in total. In February of 2010, they joined forces and set up a co operative. The idea was to counter the pervasive trend among younger Indians to leave the jungle for the suburbs of big metropoles like Manaus in search of jobs – a life mostly confined to poverty and ill health. So far, the plan has worked out tremendously. As the Amazon Charitable Trust reports, 'under the Association the community successfully built a school, a health post, a fresh-water supply and connected to the world with a solar-powered satellite internet connection. New economic opportunities like eco-tourism, film-hosting, forest product enterprises were introduced and traditional "maloca" huts for eco-tourists were constructed, all helping create sustainable livelihoods within the rainforest. The local people all own the land in the reserve jointly and share the proceeds from economic activity. This means that they all benefit from protection of the forest and defend it together, as key to their families' local culture, education, economic and physical health.' Gently, I nudge the images aside in my mind and return to Kwi, who has turned to speaking about the recordings.

'Concerning river dolphins, our guide was particularly knowledgeable about their habits and whereabouts; unfortunately my basically non-existent Portuguese limited our communication. We were taken to several dolphin-frequented places to record. Unless the dolphins emerged from the water to exhale, it was impossible to spot them. On the last night of our trip, I was recording frogs, insects and bats from the canoe. We could hear a few dolphins emerging with loud snorting sounds around us. I was listening through headphones; each exhalation sounded almost like a human's sigh. It was for me a particularly emotional experience. A five-minute extract from this recording is included on the LP.'

'I was moved by the intimacy of these recordings', I say. 'It almost seems as though they were completely unaware of your presence.'

'They might have been. Several nights I recorded from a small floating wooden platform attached to the shore. My hydrophone was about 10-12 metres deep. I was listening to the active chatter underwater: crackling, clicking, croaking, grunting, snapping and occasionally some more vocal sounds. It was very strange and alien, like listening to some emissions from deep space or radio interferences – I loved it. I spent many hours in complete awe sitting in the darkness, only sometimes opening my torch and looking around for possible creepy-crawlies. One evening I saw a hairy spider the size of my palm gliding on the surface of the water – probably some species of tarantula. I didn't know such a big creature could walk on the water! Another night, when I was in a dreamlike state, almost falling to sleep while listening to the subaquatic symphony, I heard a voice saying very loudly in my ears: "Bubak!" and then just a few splashes of water suggesting the culprit swimming away. I was startled. "Bubak" is the equivalent in Czech to "bugaboo" – indeed a rather ghostly surprise ... I still have no idea what creature it was. It might have been the boto, regarding his mischievous reputation.'

'What about the recordings realised at the Mamori, with its white water habitat?'

'The trip from Manaus to Mamori Lago is about three to four hours via a combination of minivan and smaller boats. I travelled together with Francisco López and the groups of participants of the Mamori Sound Project. Underwater, we recorded in groups, either from two to three canoes or anywhere suitable from the shore. Often you could hear active

insects in shallows close to the shore; unfortunately, there was also a lot of rotting leaves and debris which created unwanted banging sounds. Once I was recording in between the branches of a large fallen tree in the shallows. I was listening to beautiful sharp clicking insects, and suddenly I hear a sort of munching and rasping sound, greatly amplified by the hollow tree. I then noticed a large catfish eating algae and probably the wood itself. The Amazon hosts many species of prehistoric-looking catfish who are able to generate a variety of loud sounds: grunting, croaking, wrrrrrrrrrrrrr-like and ssssssssssss-sounds. Sometimes we recorded inside floating grass-carpets, which covered the surface water, called "capi" by locals. One night, under a capi, I heard electronic-like sounds as something swam slowly by, a sort of Doppler effect; it sounded exactly like pure sine-wave, sometimes saw- or square-wave. It was so strange that first I thought my equipment was malfunctioning,

PHOTO: Slavek Kwi

though I could also hear clearly an abundance of insects, air- or gas bubbles released by mud, and on top of that, other interwoven layers of electronic-like vibrations, almost like as if a fish was playing an organ. After doing some research, I discovered that these mysterious tonal signals seemed to be triggered by an active electroreception field (eod) of Gymnotiform electric fish.'

Kwi has released some of these sounds on albums like *akvatikinsekt1* and *organfish*, tiny worlds of ultra-refined sound spaces, almost impenetrable in their otherness – even to Kwi, who indirectly admitted in the press release that he was left with more questions than answers afterwards. I make a note to re-listen to them once I'm back home, then re-join Slavek in his account of his journey.

'Boto and tucuxi we encountered mainly during the trip on the double-decker boat on the river Yuma. We recorded from canoes day and night. You could hear an abundance of soniferous fish (croaking, grunting, woodpecker-like sounds), high-pitched insects and an audible spectrum of dolphin's echolocation' – he makes a trrrrrrrrrrr-like sound – 'clicks with occasional loud chuckling songs coming unexpectedly in, chasing schools of little fish jumping out of the water with loud circular splashes ... I remember not wanting to press the "stop" button on my recorder and wishing to carry on forever, to listen without a move, in complete darkness, to this ever-changing mysterious subaquatic soundscape. I was very happy there.'

'Part of the LP contains what you've referred to as "vocabulary samples". After spending a lot of time with the dolphins, what, would you say, is communicated through their sounds?'

'Vocabulary – it can be understood as "index" as well – refers to a series of short vocal sounds, which I assumed to be produced by dolphins, though this might not be completely true. Some fish, especially catfish, may generate similar sounds; the range of vocal sounds of the boto is not really known. Some calls – a sort of chuckling – sound completely the same from Mamori and from Xixuaú Xiparina. The use of sonar – echolocation - is for orientation and to locate prey in murky water. What dolphins communicate to each other – aside from territorial and mating calls – I don't know. I think it's impossible to comprehend non-human psychology; all speculation seems to me somehow inappropriate and too anthropomorphic. However, there has been some interesting research concerning "sono-pictorial" communication with sea dolphins.'

'Although you're recording animal sounds, you still sign them with your own name. What's your take on the question of authorship – shouldn't the animals be credited as well?'

'Well, if you put it that way – it seems suddenly unethical not to mention the animals. However, all this is clearly a very anthropomorphic perspective – even the scientific name for it is given by humans ... Plus, recording is always subjective', according to Slavek, 'the way it is recorded, the selection of time in and out – press record and stop – technical specifications of the microphones and recorder and, of course, decisions about editing and reproduction – all that is a very personal and creative matter.'

'This reminds me of what Eisuke Yanagisawa told me', I say, 'who likes to explore field recording by positioning his microphones in creative and unique ways, depending on the characteristics of each particular recording site. His aim, as he puts it, is to reveal the hidden beauty of the environment. But it is, of course, always his personal perspective on this beauty.'

'It feels fine to sign it with your name', Kwi agrees. 'Acknowledgement of the recording situation and location seems respectful. Animal calls are part of the acoustic environment – the same as human; in principle how does one see the question of authorship when it comes to recording human activities within the environment – such as distant hum of conversations, sounds of steps, driving cars et cetera?'

'What were your most important conclusions from the project?' I ask.

'My conclusion, after finalising *Boto*, was that the way the mind of an artist and that of a scientist operate are diametrically opposed. Art and science seem to stand beside each other in a disconnected though complementary way, creating a paradox. It seems to me impossible to combine art and science in a satisfactory way, without accepting this paradox.'

I deliberate this in my mind for a second. I am reminded of what Yannick Dauby told me: That asking scientists about animal music will most likely only either reinforce the mechanical perspective à la Descartes or anthropomorphism plus ethnocentrism on the other hand. If they are right, it could close the lid on the debate about animal music in a sense: We can keep gathering information and expand our insights into the mechanisms of sound production in animals, the argument seems to go. But at the end of the day, all this data simply relates to a different category. What we call music is precisely so attractive to us because it offers something completely different from science, because it defies the grasp of our intellectual faculties. My own research, meanwhile, which I've summed up in an article, seems to point to a different conclusion: That science and the arts seem to be converging in many respects. Kwi is sceptical: 'To converge means to come together towards a common point. There seems no other common point than observation of the same phenomenon from different angles. Though there seems to be evidence of various "combinations" of art and science, still the way the data is processed by an artist and the scientist seems fundamentally opposed. This doesn't exclude the communication in between the two modes of experience.' Still, he's curious. 'Read the article to me', he suggests. I pull a stash of papers from my bag, take a sip of water and then begin.

'To some, there can not even be the slightest doubt that animals are at least capable of musicality, even if not of creating and understanding music in the closely defined, human meaning of the term. Forgetting about Youtube videos showing cat Nora dreamily brushing the keys of a piano with her paws, dog Tucker howling along in a charmingly atonal idiom to his own improvisations and world-famous cockatoo Snowball rocking along to the beats of the Backstreet Boys for a moment, the evidence in favour of this claim seems overwhelming: Whale songs, made up of song structures, clearly identifiable phrases and aspects of rhythm, are one frequently quoted example. The creation of constant patterns and a seemingly endless amount of variations on them by birds are another. Harpist Alianna Boone has even recorded an album called *Harp Music to Soothe the Savage Beast*, a transcription of human healing-music for animal ears. The powers of the release were tested on "recently hospitalised canines at a Florida veterinary clinic" in sessions lasting several hours, and, according to Boone, "immediately began to lower heart rate, anxiety, and respiration in many cases". To the more rationally inclined, these examples were mostly outgrowths of the '60s hippie movement and could hardly be taken for much more than a novelty. And yet, more scientific support seemed to come in early 2013, when the study "Spontaneous synchronization of arm motion between Japanese macaques" by Nagasaka et al. asserted that the monkeys "showed synchronisation'" and that the results indicate "that non-human animals can establish coordination without explicit training". The first reactions to the study were ecstatic, as it seemed to suggest that although not all animals might actually be producing sounds which qualified as music, they did possess at least some of the basic cognitive traits for what is generally referred to as "musicality" – with Henkjan Honing, a professor in Music Cognition at the University of Amsterdam, Netherlands, defining musicality "as a natural, spontaneously developing trait based on and constrained by our cognitive system, and music as a social and cultural construct based on that very musicality." Animals might not always be making music, this line of reasoning went, roughly – but if so, this was a question of choice.

The hopes of those who stood by the animal music hypothesis were quickly dashed by the fact, however, that the actual contents of the Nagasaka paper did not seem to match its conclusion. For their studies, the Japanese scientists had trained monkeys in rhythmically pressing two buttons in front of them. The set-up included both a solo situation, where a monkey would be observed when alone, as well as paired with one of his companions. While the former could be said to evaluate the ability for the concept of rhythm as such, the latter offered insights into the concept of synchronicity, i.e. the locking in of individual voices into some kind of rhythmical grid shared by the group as a whole. Honing, however, wasn't convinced: "The study reports that only one of the three chimps participating in the experiment was able to do the task: a chimp named Ai. Furthermore, Ai was only able to synchronize with stimuli at a rate of 600 ms (and not at rates of 400 or 500 ms). In addition, Ai did this in reaction (positive asynchrony) and not in anticipation of the beat (negative asynchrony)", he wrote in his Music

Matters blog, "This is similar to what has been found in studies with macaques (Zarco et al., 2009; Konoike et al., 2012) that also seem to opt for a strategy of reacting to instead of anticipating a regular beat. All this, in contrast with humans that can intentionally synchronise their tapping to various rates (ranging roughly from 200 ms to 1800 ms) of a varying rhythmic stimulus (and not simply a metronome) while showing a negative synchronisation error, i.e. in anticipation of the beat.

Another point of a more methodological nature is that the experimenters used, next to sound, what they called "light navigation", a visual cue for the chimps to "remind them" of which key to press. While the authors write "it was unlikely that the visual stimuli affected tapping rhythm by chimpanzees and we can not be sure this is evidence for rhythmic entrainment in the auditory domain." The estimate by Honing is particularly relevant, because he is one of the long-term authorities on the subject, with a blog dating back to 2006. His work deals with a branch of neuroscience called cognitive science, which may well produce the most vital insights into the nature of animal music over the next decade, because it systematically analyses not just the question of music and musicality as such, but also our own ability to discern them in other species. As he once convincingly put it: "We must be careful in calling birdsong or a chimpanzee's drumming on an empty barrel, music. We make this mistake more often. We, the human listeners, perceive the sounds made by songbirds, whales, or chimpanzees as music. Whether these other animals also do that is unclear. And that makes a world of difference." And there's another issue inherent to the question. To understand the concept of animal song, we first need absolute clarity about our own concept of music in the first place.'

'Perhaps the problem is conversely that we're still too busy trying to figure out the origins of *human* music to make any truly meaningful statement about animals', Slavek proffers.

'Perhaps. And yet, we've come a long way', I say, then continue reading. 'A thesis postulated by language theorist Steven Pinker injected the debate with fresh momentum. From Pinker's point of view, music did not develop in humans from evolutionary necessity or as part of the adaptionist and non-adaptionist theories. Instead, he argued, it was the pleasant, yet ultimately superfluous side product of other, vital, capacities such as language. One could compare this, he argued, with our appetite for sweet and fatty food. At the time our ancient forebears lived, these two components – sugar and fat – were extremely rare and only made up a tiny fraction of our diet, which is why there was no point in limiting our body's craving for them. Today, at a time when they have become all but ubiquitous, the consequences can seem devastating. To Pinker, music was "auditory cheesecake" without any kind of biological purpose. It developed in sync with our own evolution as a species and now it has become as ubiquitous as sugar and fat, we can no longer get enough of it.

The avalanche of dissenting voices proved that he had hit a delicate spot. Joseph Carroll did admit, generally speaking, that "Pinker offers a splendidly fluent and lucid survey of evolutionary psychology" but asserted that "to illustrate and decorate his text, he has collected a substantial number of relevant quotations from literature, but there is no evidence that his familiarity with most of the works he quotes extends very far beyond the quotations. His literary taste and judgement seem those of an undergraduate who is extraordinarily bright but who is much more sensitive to computers than to poems, plays, or novels." Philip Ball, author of *The Music Instinct*, claimed that while "there are already strong reasons to suspect that Pinker may be wrong about the evolutionary role of music, it's a mistake to think that the fundamental value of music rests on such a disproof. The fact is that it is meaningless to imagine a culture that has no music, because music is an inevitable product of human intelligence, regardless of whether or not that is a genetic inheritance. (...) Even if Pinker were correct that music serves no adaptive purpose, you could not eliminate it from our cultures without changing our brains." Some commentators have been even less complimentary. In an interview, Pinker himself expressed the suspicion that the reason why his ideas were so passionately disliked was that they were between two stools, both rejecting evolutionary models, while also justifying them. "Everybody wants music to be an adaptation", as he put it – "and that was precisely what he did not give them."

'Pinker's cheesecake hypothesis doesn't address the topic of animal music directly', Slavek interjects.

'No, but it has undeniably served as a catalyst for the topic of music cognition in general and has thereby had an indirect acceleration effect – especially after receiving a recent boost, when a study suggested that music could arouse sensations of euphoria and the release of dopamine just like chocolate or other, tangible rewards could – an implicit hint that Pinker may have been right all along. To understand the importance of his contribution, I recommend arguably one of the most important studies on the subject, *Music, Biological Evolution, and the Brain* by Aniruddh Patel, a senior fellow at the Neurosciences Institute in San Diego, probably best known for his work with Snowball, the dancing cockatoo from popular Youtube clips. The study, freely accessible online, presents a congenial overview of his decades-long work aimed

PHOTO: Slavek Kwi

at understanding the nature of music. From the evidence he's collected, Patel concludes that music is not the result of adaptation, but of humans putting existing brain-functions originally developed for entirely different purposes – most of them related to language processing – to a new use. So far, this is still in accordance with Pinker's perspective. Contrary to Pinker, meanwhile, Patel believes that this invention was not just to satisfy a craving for a sonic sweetness, but did serve important biological purposes. His theory is that music is an invention – a purely *human* invention, I should add – which we are holding on to because of obvious benefits to our lives and well-being as a species.'

The power of Patel's reasoning is that he cuts through the brushwood of seemingly convincing, yet ultimately flimsy logic that has mushroomed around the topic. Many, for example, have claimed that because music is ancient and universal, it must by default be an evolutionary adaptation. Patel counters this by comparing the potential development of music with the control of fire. Clearly, our use of fire is not an adaptation, but an invention. And still, because of its obvious universal advantages ('the ability to cook food, keep warm, and see in dark places'), it has turned into a universal. Music, according to Patel, offers similar universal advantages, namely emotional power ('a complex emotional experience that can differ from our ordinary day-to-day emotion'), ritual efficacy ('music provides a very useful framework for certain types of rituals, independent of the emotional impact of the music per se') as well as mnemonic efficacy ('for storing long sequences of linguistic information, especially when written language is not available' – which also explains its use in ancient times as a means of transmitting and preserving information). Insights gained in musical education for children also seem to suggest that early musical training can result in disproportionally fast IQ gains – thus hinting at the idea that music has an impact on non-musical brain areas. Summing up, Patel writes: 'The example of fire-making teaches us that when we see a universal and ancient human trait, we cannot simply assume that it has been a direct target of natural selection.'

Patel then goes on to disprove the assumption that evolutionary models and invention-based theories exclude each other by default.

Language, for example, is today held to be an invention which, again because of its inbuilt advantages for humans, facilitated evolutionary brain changes. Although it is an invention, music may have led to the creation of specific brain centres dealing directly with the processing of musical information. Which may make it look like a natural result of our biology and thus, evolution. This kind of logical back and forth is characteristic of the current state of neuroscience regarding music – no wonder many are scared off by how complicated the once-charming topic has become.

Patel's discussion of two vital aspects of musicality – tonality and musical beat perception and synchronisation (BPS) – is even more revealing. As he points out, both appear to be the result of other, non-music-specific brain functions. In the case of tonality, these are related to language processing. With BPS, which is the ability to recognise beat patterns and to adapt to them, there is very strong evidence that it is based on vocal learning. The latter especially is crucial for the question of whether animal song can be characterised as music. What Patel suggests is that BPS can only take place in those species who have, mainly through gradual evolutionary changes in their brain structure, acquired the ability for vocal learning. This includes humans, very few mammals and some groups of birds. Those animals who do not possess this ability are incapable of BPS – and thus lack a vital element necessary to music's overall concept. Monkeys are not, even after a year of training them six days a week and several hours per day, able to tap their fingers along to the clicking of a metronome. On the other hand Youtube star Snowball, equipped with precisely the required capacity for vocal learning, was able to clearly demonstrate his appreciation of the Backstreet Boys. These facts support Patel's thesis: If our brain and our senses really determine what we are capable of, then, it seems, most animal song does not qualify as music.

'Isn't this more a question of definitions?', asks Kwi. 'If we take a wider view of what constitutes musicality, the results would be entirely different.'

'Absolutely. As Honing puts it: 'While it is not uncommon to see certain cognitive functions as typically human (such as language), it could well be that there are more species than just humans that have the proper predispositions for music to emerge, species that share with us one or more basic mechanisms that make up musicality. The mere fact that music did not emerge in some species is no evidence that the trait of musicality is absent.' An article in *Science* magazine from 2000/2001 also offered a dissenting perspective to Patel's. For one, it asserted that music was a lot more important than many might assume – 'The musical instruments [created 57,000 years ago]' according to one of the authors, Jelle Atema 'were more complex than the hunting tools.' Also, the authors argue that many animals use the same musical vocabulary and grammar that we do and that there is thus such a thing as a universal music. Of course, as they openly admit, whether or not this really amounts to music for the animals remains open to debate for as long as their own perspective can not be satisfyingly assessed. It's all a question of cognition, therefore – which brings us back to the beginning.'

'Perhaps you are right and the path of science may indeed bring us closer to a solution, especially where it intersects with a creative perspective', Slavek says, 'But I doubt it will actually fully deliver it.' He proceeds to the kitchen to prepare some tea and I put the *Boto* LP on his turntable, listening once more to the extreme richness of sounds, from crisp clicks panning from left to right in the stereo image to consoling creaking noises as well as the lapping of waves against the boat's hull. When I'm hearing this, concrete images inevitably enter my mind – images of dolphins lying tranquilly in the water or joyfully shooting through it, of hazy evenings spent dreaming in a hammock and hot days drifting through the jungle. What I've asked myself numerous times is whether I'm seeing all this, because the cover of the album gives some of the secret away and because I know about the background to the recordings? Or because these sounds genuinely contain some information that will infallibly spark a certain kind of visual stimulus.

'When you are listening to the piece, how does that work for you?', I call over to Slavek, who is still busy in the kitchen. 'Does it remind you of the time spent in Brazil? More importantly, is it more of an audio documentation to you or a fully-fledged composition, a through-composed work of music?'

Slavek returns and puts down two cups between us, the water steaming.

'Yes, of course it reminds me about my trips to Brazil – that is almost inevitable', Slavek laughs and then continues: 'I am treating all sounds as pure abstraction, though simultaneously I am aware of their origins. Human voices, animal calls, and other very iconic or recognisable sounds from our environment have the potential to be abstract. For example, when I hear the sound of water, I am not hearing "the music" but the water – in the moment when I hear only "the music", then I am truly listening. The training to recognise sounds as references can be overpowering. However, we have the ability to listen to everything as pure abstraction. This will again place us within a paradox.'

'It may be particularly hard, because some of these references were shaped a very long time ago. To understand the human concept of music and how it relates to the animal concepts of producing sound, I suppose we must go back in time and uncover how this concept was born and how it evolved from its origins into its highly evolved current state.'

'Yes, we must go deeper', Slavek says.

We spend the next two hours exchanging ideas and information about the history of human music and what it means to our understanding of the world around us.

Put simply, there are two fundamental concepts of the relationship between music and nature. The Eastern one assumes that human music is an interpretation of nature and, as such, part of it. The Western one sees it as being inspired by nature and therefore, inherently outside of it. It is most likely that the dominance of the Western model over the past centuries has made us focus on the *differences* between human and non-human species, rather than our *similarities*. In their analysis 'Imitation of animal sound patterns in Serbian folk music', Milena Petrovic and Nenad Ljubinkovic take an alternative perspective: 'In animal "cultures", much acoustic behaviour can be considered "musical", and in humans, music-like behaviour could have existed before the development of speech. In more than one instance, humans and animals share many musical traits and para-musical forms of behaviour: specific calls of vervet monkeys can sound like words, while male and female gibbons sing duets. Therefore, it is possible to assume that the mutual ancestors of humans and apes between 8 and 5 million years ago also had a similar repertoire of calls, and that these were the evolutionary precursor to modern human language and music'. The use of the vague term 'it is possible to assume' already indicates the ultimately speculative nature of these considerations. And yet, there can hardly be any serious doubts that, at least for many thousands of years, the development of human music was closely related to animal vocalisations through imitation and mimicry. The Tuvans living in Southern Siberia even claim that 'our music all began from imitating the sounds of animals'. Also, again according to Petrovic and Ljubinkovic, 'the early Chinese believed that humans transformed raw sounds of animals and so created music.

Today, Suya people learn songs that reproduce animal sounds. We can speculate that imitating animal sounds may have been present in the earliest human music, and may even have been its origin. It might have been that animal sounds were directly imitated for practical purposes or used as ritualised imitation for magical purposes.' These and similar theories have recently been supported by scientific research into the reasons why music affects physical and psychosocial responses, which has drawn parallels between the 'innate' musical capacities of non-human primates and the language capacities of humans.

In some cases, the imitation of animal sounds might also have served a far more mundane purpose, as Emily Doolittle, author of *An Investigation of the Relationship between Human Music and Animal Songs* (Princeton, 2007), stresses – entertainment: 'Inuit throat singers perform songs that imitate the sounds of sled dogs and geese indoors for their own amusement, where contact with the animals in question isn't even possible. Indigenous Bolivians sing songs that consist of a great number of birdsong imitations, joined together over a simple musical ostinato, also for their own pleasure. An Alan Lomax recording from the 1950s documents Annie Johnston from the Isle of Barra (Scotland) singing in imitation of a collection of local birds, including grouse, black-backed and ordinary crows, rooks, chickens, puffins, razorbills, various thrushes, and seagulls. The songs use nonsense syllables to imitate the birdcalls, and also tell little stories about what the birds might be doing. In similar spirit (though without the accompanying stories) are a series of wax cylinder recordings and later records, made beginning in 1906 and continuing until the 1930s, of American Charles Crawford Gorst whistling local birdcalls.'

Mimicking animal sounds has remained pervasive in many cultures until today. The Nyae Nyae community in the Kalahari desert, for example, will imitate animal rhythms in some of their games – the task of the participants is to present them as accurately as possible to allow for the others to determine the animals. Perhaps, recognising the deeper meaning of these communications, she suggests, one could regard these activities as a form of engaging with the landscape – and therefore as an example for the Eastern concept of music, of seeing oneself as an integral part of the biophony and contributing to it. It is easy to see how our ancestors would then go from mere imitation to a more abstracted art form and to sounds, which were still based on imitations, but which transformed them to a degree which would make it impossible for someone unfamiliar with them to determine their origins by ear.

Fascinatingly, the Tuvans don't just have a single word for 'music', but instead, according to Doolittle, 'have several words that distinguish between the degrees of pure imitation and musicalisation in a sung or played work. "Ang-meng mal-magan öttüneri," for example, refers to purely imitative sung pieces, some of them undistinguishable from their natural sources to the untrained ear. Animals imitated include a variety of wild and domestic birds, horses, camels, snakes, bulls, and wild boars. "Xöömei" refers to throat-singing, initially developed as an aestheticised form of imitation of natural sounds. Xöömei blends seamlessly with the sounds of the natural environment, but is immediately recognisable as human in origin.'

It is very likely that at some point, during the gradual development of music from its imitative functions to an art form, the singers became aware of the fact that their abstractions were capable of evoking a greater emotional potential than more concrete sounds. Composer

Michael Pisaro has convincingly argued that a pitched note (i.e. a sound in the part of the spectrum we recognise as 'musical') is a very rare occurrence in nature – and for exactly that reason, it is a very powerful tool for expression. The invisibility of music, its emotional potential and spiritual qualities, gradually transformed it into a language reserved for communicating with the gods – which, in an interesting parallel, in many cases, took on the likeness of animals. In the Christian world, the strong emotional potential of music aroused suspicion and distrust. And so, in a bid to distance oneself from it, these effects were transferred to animals, which explains why there are so many mediaeval descriptions of animal orchestras or animals joining in human performances: by conferring the arousing powers of sound to other species, it was thought, one could analyse, tame and control them.

As David Whitwell points out in *Essays on the Origins of Western Music*: 'In reading ancient literature one is struck by the frequent reference to animals when the subject under discussion is music. The range of animals is worthy of a zoo. In addition to dolphins, one reads of swordfish, elephants, horses, oxen, lions, tigers, bears, deer, wolves, asses, hogs, sloths, dogs, cats, silk-worms, flies, frogs, snakes, birds of many kinds, dragons and often just "wild animals." Frequently when we read of animals, it seems we are really reading about people. On the other hand, sometimes we find references in which the musical attributes are really those of the animal itself, as in the case of Kircher's 17th century account of the Central American sloth: "It perfectly intones as learners do, the first elements of music, do, re, me, fa, sol, la, sol, fa, me, re, do. Ascending and descending through the common intervals of the six degrees, insomuch that the Spaniards, when they first took possession of these coasts, and perceived such a kind of vociferation in the night, thought they heard men accustomed to the rules of music." Kircher concluded, "If music were first invented in America, I would say that it must have begun with the amazing voice of this animal."'

As Whitwell points out, this train of thought eventually led to music not so much as a form of engaging with the world, but of controlling it. In the 6th century, Roman statesman and author Cassiodorus discovered that the Hypodorian mode was best for calming down wild beasts. In the 16th century, in an interesting twist to the so-called Mozart effect, Giustiniani claimed that silk worms became more productive when listening to singing and playing. Consequently, practical purposes took over. Imitating animals could have served to attract some for hunting and scaring off others that might represent a potential danger. According to Doolittle, 'the peoples of Mongolia, for example, who have a long and continuous history of herding such animals as sheep, cows, goats, horses, camels, and yaks, imitate and communicate with animals through a variety of means, including song, whistling, overtone singing, and playing instruments. They sing syllables derived from the sounds of ewes, goats, and cows respectively, to prevent mothers from rejecting their young, and mare-like syllables to encourage mares to give milk. They use voices and instruments to imitate such animals as mountain goats and wolves in hunting as well', while 'traditional Norwegian herders sing imitative songs to communicate with their goats' and 'Swedish herders play pieces to their cows on three- or four holed instruments made of cow or ram horns, and on birch-bark bugles up to two meters long, both of which have a timbre not unlike the lowing of cattle.'

It can safely be assumed as well that imitation was used to communicate with the first domesticated animals. This would later lead to a more elaborate employment of music and sounds to control big herds of animals. Polybus, in the 2nd Century BC, expresses his admiration, for example, for the herders of Corsica and Italy, who were capable of assembling and leading their animals with a horn. In Corsica, this seemed particularly vital, as the rocky landscape meant that herds were usually dispersed, with individual goats spread out across the mountains like wild animals. And yet, to Polybus' amazement, as the shepherd 'sounds his horn, the herd will run off at full speed and gather round the horn.' In Italy, meanwhile, where herds of swine could be particularly big and easily mixed with others grazing on the same land, the horns of different herders were capable of separating the groups from each other with perfect ease and precision, where eyesight would have been unable to perform the same task.

'So, one could say, summing up, that human music first developed in response and in parallel to animal sounds. Then, through playful transformation, it gradually turned into something decidedly more complex', Slavek says. 'Whether or not this disqualifies it as "music" isn't entirely clear – it seems, in fact, to have been a point of contention for centuries.'

'Which is why we should perhaps turn towards an important aspect which hasn't been discussed thus far: The topic of consciousness and aesthetics', I say.

'What do you mean?'

I lean back in my chair, sipping my now-cold tea. 'Well, many seem to consider it vital in terms of classifying sounds as music whether or not they were created with an underlying awareness of aesthetics. Spanish sound artist and field recordist Francisco López, for example, noted that an animal could produce, perceive, transform and play with a sound only to improve its communication skills. Animals did that when they were learning, according to López, so their reasons for producing sound could be something besides aesthetics. At the same time, I've spoken to quite a few other artists about this and all of them have questioned the validity of his statement. For someone like Geoff Sample, it is perfectly clear that if a starling picks up on a car alarm and then proceeds to copy its sound – sometimes even with slight personal alterations or variations – it is making aesthetic choices. If you're copying a noise and then modifying it, you're doing so because the modified sound is more appealing to you.

The same goes for the discovery of new calls: Scientists agree that accidents and mistakes play a huge role in the extension of a bird's vocabulary. And yet, whether or not a particular new element is retained within its repertoire or not is a question of choice – and, therefore, of aesthetics. This thought is similar to what Rodolphe Alexis told me: "Aesthetics seem to be about a sense of perfection. They're closely linked to a particular, desirable function, such as when the great apes are manufacturing tools to pick up termites. It is rather close to what Aristotle meant with the 'techne', as opposed to the 'praxis' which applies solely to humans ..." It also seems as though, as Yannick Dauby has pointed out, some birds, like parrots and ravens, actually produce sounds just for themselves and not for their interaction with others. So communication doesn't need to be a factor. And, as Alexis has rightly stressed, the question of consciousness is complicated anyway. Humans, too, are constantly influenced by subliminal factors like magnetic fields and we communicate with our environment through electrochemical activity, without actually noticing it. The sounds of nature, as he puts it, are like a stream running through us. How, then, can we question the consciousness of other species, if we can't even be sure about our own?'

Slavek nods in agreement: 'It is a mystery!'

'To me personally, the real mystery is why so many are still doubting the possibility of aesthetic judgement in animals. Let me give you a quote by Dauby when I asked him this very question: "The existence or not of an animal aesthetic has a philosophical importance and might change our whole consideration for animals in general. But I think we're having difficulties to define or discuss it because we're lacking descriptions for the situations where animal invention takes place. Also, the experiments which have been made are probably very naive. The topic is wide open and it is a territory made for collaboration between science and art. Unfortunately, the few examples of human and non-human musical interactions are often limited and extremely stereotyped. I think we're simply asking the wrong questions." He also had another interesting point, which sheds some light into how hard it is to arrive at definitive statements. If we research the question of aesthetics, do we analyse the behaviour of an "average" individual of a particular species or of a "special" one – i.e. one with particularly highly developed "musical" facilities? If we choose the former, we risk making broad-sweeping statements – just like we would if we were to determine the musical facilities of humans by looking at the "average" non-musician. If we choose the latter, we risk turning exceptions into the rule. It would seem that there is no escape from this dilemma.'

It is long beyond dusk now and we light a few candles around the room to avoid lapsing into complete darkness. Strange shapes and figures are dancing on the walls and the ceiling, almost like in an ancient cave.

'Horse trainer Linda Kohanov has pointed out the relevance of pre-verbal communication – the idea that we may have communicated with animals before we were even able to speak ourselves. The importance of the pre-verbal level may explain why we feel this strong connection with animal sounds and why we already felt it at a stage when human sound production was still at a level many today wouldn't even consider music at all – the stage of imitation we just spoke about', I say. 'This is a pure love for sound, and it connects us with a time when this pure love was our main connection with the animal world, when we weren't thinking in concepts so much.'

'Didn't you speak to some of the field recordists about this when you were doing the interviews with them?'

'Yes I did.'

'So what did they say?'

'Mike Webster of the Macaulay Library told me something very interesting: The main purpose of the Library, according to him, has always been to provide data for research and to approach the recordings in a highly organised, scientific way. And yet, many people actually come to their website just to listen – because they *enjoy* listening, because it makes them feel this very connection I just spoke about. His interpretation of this phenomenon is, and this runs counter to the general opinion of humans experiencing the world mainly through the eye, that we are "auditory animals". As Mike put it, "deep in our evolutionary history is an appreciation for sound, for 'music' if you will. Hearing the sounds of nature, and of music, stirs deep feelings. Natural selection has fixed this appreciation for the sounds of nature in our DNA."'

I also spoke to bird specialist Geoff Sample and he explained that there were many different ways in which animal calls could leave an impression on him. Sometimes, he might be moved by the subtlety of the sounds, at other times by a passionate delivery. But not all of this, he stressed, was 'beautiful' in the way we would usually use the word and that 'even harsh or monotonous sounds, heard in context, can contribute to the beauty of a natural soundscape and help establish a dynamic narrative in the flow of a movement.' Mike Webster spoke of the Macaulay Library as creating a dictionary to help us decode what animals are communicating through their music. To Sample, the more this dictionary grows and the more we internalise it, the more we'll understand what he calls the 'behavioural ecology of a species' and how a song works within the breeding system of a species – and will trigger different, increasingly deeper emotional responses. What is so particularly moving is tapping into a communication system that has evolved over millions of years, branching out into myriads of different idioms and dialects, into a multiplicity of different forms. 'When you are familiar with all the range of sounds, it can inspire a feeling of intimacy and security in the world', he told me, 'particularly noticeable when you're out in a remote place in the darkness.'

We sit in silence for a while, thinking about this. Then I continue.

'The most essential thing he said, however, was something different. Of course, ever since the Cagean revolution, many have come to accept that everything – from silence to an indeterminate organisation of sounds – can be considered music. So what is preventing us from accepting animal sounds as music, then? According to Sample, it may be the historical and cultural context associated with a musical work. He asks a very important philosophical question to help us understand our own bias in this regard: Would the experience of listening to bird song be the same if these sounds were being produced by mechanical devices hidden in the bushes?'

Slavek leans in, excitedly, 'I completely agree. The idea of "music" is a human concept, therefore the parameters and the aesthetics of "music" seem to me inevitably changeable – fluid, rather than static – and subject to individual perceptions. In this sense, I am not concerned about the "musicality" of the sounds. We might consider sounds as "musical" when these sounds make sense to us, when it sounds meaningful to our ears. This depends solely on the development of our perception and cultural conditioning.'

'And by the same account, we can assume that the reason why so many animals are not responding to human music is not that they're

THE GRASSHOPPER AT TWILIGHT

lacking "musicality", but simply because it isn't relevant to them', I say, 'Your approach is perfectly in sync with what David Teie said, who broke down the question of "is it music or isn't it?" to a very simple equation: "If it is produced like music and sounds like music, most people would agree that it *is* music." The more our own understanding of music expands, the more likely it is that animal sounds become part of it – not as a question of quality, but purely of definition. David Velez told me a nice example to illustrate this point, one which every phonographer will instantly recognise: Say you're out recording a sewer for one of your sound art projects. Then these sounds will definitely, almost doubtlessly represent music to you. To outsiders, gazing at you with a somewhat bewildered look, they will just be ... the noises produced by a sewer. Music, as he puts it, happens in our conscience – we are creating it in our mind.'

'I like the idea that our concept of music is malleable', Slavek says, 'I am trying to create such situations which escape my own analytic perception, such situations which do not make sense, that unfold suddenly, leaving space for surprise, and freedom of interpretation.'

'Or even for mystery', I posit. 'Jez riley French told me that he doesn't have a firm answer to the question of what animals are saying, but neither would he want one. Because that would mean he would have stopped being fascinated, which, in turn, would mean he would have to stop exploring. Listening is forever an emotive process to him and one can listen to the sounds produced by animals without focussing too much on their language. Music, sound art and sound – these are loaded terms, especially if they serve to defend or justify egoistical goals. And so his own perspective is a more personal one: "I'm a father, a son, try my best to be a good person, a listener and enjoy exploring audible and visual experience. I like moments, both short and those rare elongated moments that hold you for longer durations. Those are the definitions that really matter and say something about me and how I feel about things."'

'Quite wonderful', Slavek says, 'I am also perpetually oscillating between wanting and not wanting to know – I feel like knowledge somehow limits the purity of an aural experience and simultaneously it can complement it in a strange contradictory way - "to know" and "not to know" is the paradox.' He smiles, then adds: 'So what, after all these conversations and interviews, is your own definition of music and whether or not it applies to animals?'

I think back to all the people I've spoken to, all with their unique approaches, backgrounds and perspectives. Perhaps, if one were to put a gun against my head and force me to arrive at a definition, I would say that music is *interpreted sound*, that it happens, when the events reaching the ear are arranged into some kind of order by our mind. I would probably stress the fact, just like David Velez, that music is defined less by the creator, and more by the listener and that I don't see any reason why animals shouldn't be able to make sense of the world through sound as well – but that they just happen to arrive at different conclusions. And I might also add that what we call music today is the result of thousands of years of evolution and creative development and that just because it has changed radically as part of that process, none of that devalues its roots, which happen to be our roots as human beings.

In fact, the more our perspective of music and sound puts us into connection with our own past as a species, the more we will be able to establish a connection with others. Not just through knowledge, but through our shared experience on this planet. As Jana Winderen put it: 'There is an inherent rhythm in all living, a pulse. If a fish is chewing on something, there is a rhythm. When the sea urchin is filtrating the water, there is a rhythm. There is even a rhythm just in the waves lapping, the pulse of the wind. It gives me hope to listen to the animals, to listen closely and look closely at how the creatures around us put a perspective on life. And hopefully, it can change the brutality with which we step around in this world, with our earmuffs and blindfolds on. We need to pay attention. When I just had a break now, I saw a huge grasshopper. I was so happy to see it, since I have not yet heard them this summer. I am looking forward to twilight when they start playing – best concerts ever!'

This is what it is all about, I realise suddenly, as the sun slowly rises and a mesmerising dawn chorus sets in. Music is our way of letting the world know we're still alive. That's why it means so much to us, even if we can't understand it. That's why we think it's so beautiful.

Xixuaú Xiparina Reserve http://www.amazoncharitabletrust.org/en/projects-xixuau-xiparina.asp

Maryann Mott, Pets Enjoy Healing Power of Music http://www.livescience.com/4791-pets-enjoy-healing-power-music.html

Henkjan Honing, Are monkeys capable of rhythmic entrainment? http://musiccognition.blogspot.de/2013/05/are-non-human-primates-capable-of.html

Henkjan Honing, Is birdsong music? http://musiccognition.blogspot.de/2012/10/is-birdsong-music.html

Henkjan Honing, Can the origins of music be studied at all? http://musiccognition.blogspot.de/2013/01/can-origins-of-music-be-studied-at-all.html

Henkjan Honing, 'Vocal mimicry hypothesis' falsified? [Part 2] http://musiccognition.blogspot.de/2013/05/vocal-mimicry-hypothesis-falsified-part.html

Why Music? http://www.economist.com/node/12795510

Joseph Carroll, Steven Pinker's Cheesecake For The Mind, Philosophy and Literature 22, 1998

Philip Ball, The Music Instinct, Oxford University Press, 2012

Jen Mapes, Do Animals Have an Innate Sense of Music? http://news.nationalgeographic.com/news/2001/01/0105biomusic.html

Milena Petrovic and Nenad Ljubinkovic, Imitation of animal sound patterns in Serbian folk music, journal of interdisciplinary music studies fall 2011, volume 5, issue 2, art. #11050201, pp. 101-118

David Whitwell, Animals and Music, Ancient Views of What is Music, Whitwell Books, 2014

Doolittle, Emily, 'Crickets in the Concert Hall: A History of animals in Western music'. TRANS / Transcultural Music Review 12, 2008

ANIMAL MUSIC

CD track details

ABOUT THE CD

01: TIKAL DAWN - ANDREAS BICK, GERMANY

Parrots (*Psittaciformes*), the montezuma oropendola (*Psarocolius montezuma*), a Guatemalan turkey (*Meleagris ocellata* or *Meleagris gallopavo*) and the notorious howler monkeys (*Alouatta*)

In March 2008 I travelled Mexico and Guatemala. The most spectacular Maya-site is probably Tikal in the midst of the Guatemalan jungle. I had the opportunity to stay at gran plaza in the centre of the ruins about 45 min. before the gates open at dawn. My guided group rushed to the temple with the best sight over the forest canopy, while I was alone with a lot of wildlife in the heart of this abandoned city. There are a lot of parrots, the montezuma oropendola again, that I already got acquainted with on the Cano Negro in Costa Rica, a Guatemalan turkey and – indeed – the notorious howler monkeys.

www.andreas-bick.de

02: HERMETICA - DANIEL BLINKHORN, AUSTRALIA
Hermit crabs (*Paguroidea*)

'Hermetica' attempts to capture the duality, if not vitality accompanying the life of a colony of hermit crabs. After encountering a surreptitious colony of hermit crabs on an island off the coast of Venezuela, I was astounded to discover the wonderful world of sound contained within ... Fascinated by the beautiful, arcane and hermetic textures created as the crabs jostled and wrestled over one another in a somewhat confined space, I lowered a microphone into the colony in an attempt to eavesdrop, hoping to capture some semblance of their activities. Much to my surprise, the intensified mass of beautifully articulated sound I heard produced a distinct impression of singularity as the crabs grappled and vied within the colony. As I listened, what struck me most was the disparity occurring between that which I saw, and that which I heard. To see the crabs as they moved so slowly and awkwardly over one another produced a striking contrast to the sheer density and intensified activity portrayed by the sounds this action appeared to make. The composition

presented was very carefully arranged to reveal this heightened aural activity, while nudging and shifting some of the sounds, the overall shape of the work was designed to create a short portrait of the colony. In such a way I had hoped to make a short work that brought into question how an animal seemingly as slow, and even ungainly, could produce such manifold dimensions of sonic activity, motion and dexterity. The piece forms part of a seascape triptych within the suite 'terra subfónica'

Taken from 'terra subfónica' Daniel Blinkhorn, Gruen 117

www.gruenrekorder.de/?page_id=9841

03: AMAZONS & PARROTS - RODOLPHE ALEXIS, FRANCE
'White-fronted Amazon' (*Amazona albifrons*), an 'orange-fronted parakeet' (*Aratinga canicularis*) and a perched pair of 'yellow-napped parrot' (*Amazona auropalliata*)

A group of white-fronted Amazon, an orange-fronted parakeet (call in flight) and a perched pair of 'yellow-napped parrot'. A windy morning in the dry and deciduous forest of the Natural Park of Santa Rosa, Costa Rica. Recorded 24 Nov 2011.

www.rodolphe-alexis.info

04: GRAND CANAL SPRINGS (EXCERPT) - TOM LAWRENCE, IRELAND
Waterbug / Water Scorpion (*Nepidae*)

This recording is a protracted performance given by a single waterbug (water scorpion) in heightened antagonistic stance. The hydrophone has been positioned about one inch from the insect. A reading of 110 dB was measured during this recording. The insect stridulated in this way for approximately nine hours. In the background the communicative songs of the wider ecosystem can be heard. Towards the end of the recording an interesting oscillation technique takes place.

Taken from 'Water Beetles of Pollardstown Fen' Tom Lawrence, Gruen 087

www.gruenrekorder.de/?page_id=5227

05: SEALS - MARTIN CLARKE, UNITED KINDOM

Common seal (*Phoca vitulina*)

Recorded on Mousa Isle, off the East coast of Shetland, UK, in August 2007.

www.rockscottage.net

06: BOTO (EXTRACT) - ARTIFICIAL MEMORY TRACE, IRELAND

Amazonian pink dolphins (*Inia geoffrensis*)

*Boto is the local name for Amazonian pink dolphins (*Inia geoffrensis*). Next to voices and the sonar of dolphins, you can hear barking catfish, aquatic insects and crustaceans. Recorded 27.10.2011 (night) in the river Yuma during Mamori residency in Amazonas, Brazil by Slavek Kwi. Equipment: Sound Devices 722, hydrophone CRT C55. Special thanks to Francisco López.*

www.artificialmemorytrace.com

07: ADÉLIE PENGUINS (EXCERPT) - CRAIG VEAR, UNITED KINDOM

Adélie penguins (*Pygoscelis adeliae*)

In the winter (Austral summer) of 2003/4 I embarked on an ambitious musical project in Antarctica, having been awarded a joint fellowship from Arts Council England and the British Antarctic Survey's Artists and Writers Programme. The purpose of my visit was to compile a unique library of field recordings from the Antarctic and sub-Antarctic regions, which would become the sound source for music composition. I journeyed to far and desolate lands, recorded colonies of penguins and seals, flew to isolated huts deep in the Antarctic Peninsula, and smashed through pack ice aboard an ice strengthened ship. I experienced the euphoric highs and the mind-crushing lows of solitude, the overwhelming presence of all who had come and gone, together with the realisation that I was, as a human and an artist, a mere speck on this planet.

Taken from 'Antarctica' Craig Vear, GrDl 089

www.gruenrekorder.de/?page_id=4733

08: PILOT WHALES (EXCERPT) - HEIKE VESTER, NORWAY / GERMANY
Pilot whales (*Globicephala*)

Pilot whales appear year-round in the Vestfjord. We do not know where they come from, and we always see different animals. This recording was from a resting and slow-travelling group that came close to the boat several times, producing a variety of different calls, buzzes and clicks. Location: Vestfjord, outside Skrova (2008).

Taken from 'Marine Mammals and Fish of Lofoten and Vesterålen' Recorded by oceansounds / Norway | Heike Vester, Gruen 066

www.gruenrekorder.de/?page_id=382

---

09 BRAME, SEPTEMBRE 2011 - MARC NAMBLARD, FRANCE
Red deer (*Cervus elaphus*), great green bush crickets (*Tettigonia viridissima*), wild boar (*Sus scrofa*)

*Mid-September in a forest of the Vosges region. The frenetic red deer (*Cervus elaphus*) rutting season is in full swing. We are on the edge of a thick and thorny forest, a few meters away from a large meadow. Clusters of languid deer are grazing in the darkness. Deer bellow and listen. They are very stimulated. High perched in the trees, great green bush crickets (*Tettigonia viridissima*) scatter their impassive songs. Wild boar (*Sus scrofa*) meander about.*

www.marcnamblard.fr

10: FORMICA AQUILONIA, SWEDEN - JEZ RILEY FRENCH, UNITED KINDOM
Wood ant (*Formica aquilonia*)

Recorded with 2 x JrF d-series hydrophones placed into an ant nest during a trip to Sweden in May 2012. I feel more able to get closer to the sounding world by understanding its intricacy and reality. It's powerful.

www.jezrileyfrench.co.uk

11: SCHWEBFLIEGEN - LASSE MARC RIEK, GERMANY

Hoverflies (*Syrphidae*)

Recorded directly on a plant at a forest near Alajärvi, Finland, 2007.

www.lasse-marc-riek.de

12: CENTRAL MONGOLIAN HIGH MOUNTAIN RANGE HABITAT - PATRICK FRANKE, GERMANY

Bar-headed goose (*Anser indicus*), ruddy shelduck (*Tadorna ferruginea*), saker falcon (*Falco cherrug*), red-billed chough (*Pyrrhocorax pyrrhocorax*), altai accentor (*Prunella himalayana*), brown accentor (*Prunella fulvescens*), eurasian twite (*Linaria flavirostris*), siberian marmot (*Marmota sibirica*)

04/06/2013 Bayankhongorijn Khukh Nuur
47°31'0.98"N 98°30'44.99"E Mongolia
altitude: 3,100m

www.singwarte.info

13: OTUS SPILOCEPHALUS - YANNIK DAUBY, FRANCE

The mountain scops owl (*Otus spilocephalus*)

I always have a bit of nostalgia for an owlet that I haven't heard for years: Otus scops *(Hibou Petit-Duc in French). Its soft calls are perfectly representative of the starry and perfumed nights of the Mediterranean Alps. In Taiwan, fortunately there is a related species, the commonly found* Otus spilocephalus *(黃嘴角鴞 in Chinese) that gives me the same feeling of familiarity. But here, the landscape is much more humid, in the heart of a nature reserve in the Northern part of the island, Fushan Botanical Garden (福山植物園). Around the tree where I recorded it, dozens of tiny tree frogs,* Kurixalus eiffingeri *(艾氏樹蛙) are weaving delicate patterns and a rarer owl sings in the distant hills,* Ninox scutulata *(褐鷹鴞). This recording was done on the 18.02.2013 around 9pm, during field work organised by Guandu Nature Park (關渡自然公園 http://www.gd-park.org.tw/).*

www.kalerne.net/yannickdauby

14 UNTITLED#292 - FRANCISCO LÓPEZ, SPAIN

Shearwaters (*Calonectris diomedea diomedea*)

This one I have identified! Floating in the waters of the coast of Morocco. Crazy stuff, the birds sound like crying babies in the water at night (perfect for sailors' myths! ;-)).

www.franciscolopez.net

15. SUMMER SUNSET 01 - ECKHARD KUCHENBECKER, GERMANY

Oecanthus pellucens and *Tartarogryllus burdigalensis*. *Ephippiger ephippiger, Pelophylax sp.* There are several species of 'green frogs' in the South of France with very similar calls. It think it is impossible to precisely identify them in this recording.

A valley near Escayrac, Summer 2003, after dawn. In the first week of August 2003, we visited friends living close to the city of Cahors in the South-West of France. During this week, we realised various acoustic documents of the nature of that region. I was fascinated by the silence of this remote valley, permeated by a tiny brook and an all but car-free street. The heat of the first days of August made human activity all but impossible – even noises from a nearby farm were scarce at best.

I created a TimeLapse from a 28-minute long recording. I used 15 seconds of audio, before skipping the next 45 seconds, then segueing into the next 15 again with a one-second transition and so forth. As a result, the 28-minute dawn is condensed into seven minutes, while still running at its normal speed. This means you can hear the changes taking place in the soundscape very clearly. You can hear various crickets and cicadas as well as frogs beginning to croak in the wetlands of the brook. Two airplanes coming from Toulouse are flying over the scenery. The dynamic peaks are clearly visible and probably caused by another cicada species. However, they are almost inaudible to the human ear, since their frequencies are mainly within the range of 18000-24 Khz. They seem to be caused by just two individuals, communicating with each other throughout the entire dawn.

www.your-sounds.com